AF885328

EYE ON AI MEETING NEEDS SUSTAINABLY

PHILOSOPHY OF DESIGN ACTION AND MULTIPLICITY

BOOK 5

DAN PAUL

PROACT 2525

CONTENTS

CHAPTER 1
SUSTAINABLE DEVELOPMENT

Artificial intelligence (AI) has immense potential to contribute to sustainable development across various domains. Some key goals for AI in sustainable development include:

Environmental Monitoring and Conservation: AI can be utilized for monitoring and managing environmental resources such as air, water, and forests. It can analyze vast amounts of data from satellites, sensors, and other sources to track changes in ecosystems, identify patterns of deforestation, monitor wildlife populations, and predict natural disasters.

Climate Change Mitigation and Adaptation: AI can aid in developing models to predict climate change impacts, optimize energy usage, and design resilient infrastructure. Machine learning algorithms can analyze climate data to identify trends, support climate risk assessments, and inform decision-making for adaptation strategies.

Renewable Energy Optimization: AI algorithms can optimize the generation, distribution, and consumption of renewable energy sources such as solar and wind power. By forecasting energy demand, optimizing grid management, and improving energy efficiency, AI can help accelerate the transition to clean energy systems.

Sustainable Agriculture and Food Security: AI technologies like precision agriculture, crop monitoring, and predictive analytics can enhance agricultural productivity while minimizing environmental impacts. AI-powered systems can optimize resource usage, detect pests and diseases early, and improve crop yields through personalized recommendations for farmers.

Healthcare Access and Equity: AI-enabled healthcare solutions can improve access to quality healthcare services, especially in underserved regions. Telemedicine, diagnostic tools, and predictive analytics can help diagnose diseases early, personalize treatments, and optimize healthcare resource allocation.

Education and Skills Development: AI can facilitate personalized learning experiences, adaptive tutoring systems, and digital educational content to enhance learning outcomes for individuals worldwide. By providing access to quality education and vocational training, AI can empower people with the skills needed for sustainable development.

Social Equality and Inclusion: AI technologies should be designed and deployed in a way that promotes fairness, transparency, and accountability. Efforts to mitigate biases in AI algorithms and ensure inclusive decision-making processes can help address social inequalities and promote diversity in AI development and usage.

Responsible Governance and Ethics: Ethical guidelines, regulations, and standards are essential to ensure that AI technologies are developed and deployed responsibly, respecting human rights, privacy, and ethical principles. Multistakeholder collaboration and transparent governance mechanisms can foster trust and accountability in AI systems.

By aligning AI development with sustainable development goals and leveraging its transformative potential responsibly, we can harness the power of AI to address pressing global challenges and create a more sustainable and inclusive future for all.

CHAPTER 2
SUSTAINABLE TRANSPORT

Artificial intelligence (AI) can play a significant role in advancing sustainable transport by optimizing systems, reducing emissions, improving safety, and enhancing efficiency. Here are some specific AI goals for sustainable transport:

Smart Traffic Management: AI-powered traffic management systems can optimize traffic flow, reduce congestion, and minimize travel time and fuel consumption. By analyzing real-time data from traffic cameras, sensors, and GPS devices, AI algorithms can dynamically adjust traffic signals, reroute vehicles, and manage traffic patterns to optimize road capacity and reduce emissions.

Electric and Autonomous Vehicles Integration: AI can facilitate the integration of electric vehicles (EVs) and autonomous vehicles (AVs) into transportation networks. AI algorithms can optimize charging station locations, predict EV charging demand, and manage energy usage to support widespread adoption of electric vehicles. Additionally, AI-enabled autonomous driving systems can improve road safety, reduce accidents, and optimize vehicle routing and scheduling.

Public Transportation Optimization: AI can enhance the efficiency and accessibility of public transportation systems. By analyzing passenger data, traffic patterns, and service schedules, AI algorithms can optimize public transit routes, schedules, and capacity to meet demand while minimizing energy consumption and emissions. AI-powered predictive maintenance systems can also help prevent breakdowns and reduce service disruptions.

Shared Mobility Services: AI-driven shared mobility services, such as ride-sharing, car-sharing, and bike-sharing, can promote sustainable transportation options and reduce the need for private car ownership. AI algorithms can optimize fleet management, match passengers with vehicles, and dynamically adjust pricing and incentives to encourage shared rides and reduce empty vehicle miles.

Infrastructure Planning and Design: AI can support sustainable infrastructure planning and design by analyzing transportation data, urban development patterns, and environmental factors. AI algorithms can optimize the design of roads, bridges, and public transportation networks to minimize environmental impact, improve accessibility, and enhance resilience to climate change.

Emissions Monitoring and Reduction: AI can help monitor and reduce greenhouse gas emissions from transportation activities. AI-powered systems can analyze vehicle emissions data, identify emission hotspots, and develop strategies to reduce emissions through vehicle electrification, fuel efficiency improvements, and alternative transportation modes.

User-Centric Mobility Solutions: AI can personalize mobility services and experiences to meet the diverse needs of travelers. AI-driven mobility platforms can provide real-time information, trip planning assistance, and personalized recommendations based on user preferences, mobility patterns, and environmental goals.

By leveraging AI technologies to address these goals, transportation systems can become more sustainable, efficient, and resilient, contributing to reduced emissions, improved air quality, enhanced safety, and better mobility options for people and goods.

CHAPTER 3
MEETING NEEDS LOCALLY

Artificial intelligence (AI) can be instrumental in meeting local needs efficiently and sustainably by addressing various challenges and enhancing community resilience. Here are some AI goals for meeting needs locally:

Community-Based Decision Making: AI can support community-based decision-making processes by analyzing local data, preferences, and feedback from residents. AI-powered systems can help identify community priorities, allocate resources effectively, and facilitate participatory decision-making to address local needs and challenges.

Hyperlocal Services and Solutions: AI can enable the development of hyperlocal services and solutions tailored to the specific needs of local communities. By analyzing local data and preferences, AI algorithms can personalize services such as healthcare, education, transportation, and public safety to meet the unique needs of residents in different neighborhoods.

Local Economic Development: AI can support local economic development initiatives by identifying opportunities for entrepreneurship, innovation, and job creation. AI-powered analytics can analyze local market trends, consumer behavior, and business dynamics to inform economic development strategies and support the growth of local businesses and industries.

Resilient Infrastructure and Services: AI can enhance the resilience of local infrastructure and essential services against natural disasters, climate change, and other disruptions. AI-powered predictive analytics can assess risks, optimize emergency response plans, and improve the reliability and efficiency of critical infrastructure such as water systems, energy grids, and transportation networks.

Environmental Monitoring and Management: AI can contribute to local environmental monitoring and management efforts by analyzing data from sensors, satellites, and other sources. AI algorithms can detect environmental hazards, track pollution levels, and provide early warnings for natural disasters, helping communities mitigate risks and protect natural resources.

Equitable Access to Services: AI can help ensure equitable access to essential services such as healthcare, education, and transportation for all members of the community, including vulnerable populations. AI-powered solutions can identify barriers to access, optimize service delivery, and tailor interventions to address the specific needs of underserved communities.

Community Health and Wellbeing: AI can support community health and wellbeing initiatives by analyzing health data, predicting disease outbreaks, and optimizing healthcare services. AI-powered tools can facilitate early detection of health risks, personalized treatment plans, and proactive interventions to improve health outcomes and quality of life for residents.

Cultural Preservation and Heritage: AI can aid in preserving and promoting local cultural heritage and traditions through digitization, preservation, and dissemination of cultural artifacts and knowledge. AI-powered platforms can facilitate community engagement, storytelling, and cultural exchange, preserving local identities and fostering social cohesion.

By focusing on these AI goals, local communities can harness the power of technology to address their unique needs, strengthen social bonds, and build sustainable and resilient societies.

CHAPTER 4
MEETING NEEDS SUSTAINABLY

Meeting needs sustainably with the help of artificial intelligence (AI) involves leveraging technology to address various challenges while minimizing negative impacts on the environment and promoting long-term viability. Here are some AI goals for meeting needs sustainably:

Resource Optimization: AI can optimize resource usage across various sectors such as energy, water, and materials. By analyzing data and predicting demand, AI algorithms can optimize resource allocation, reduce waste, and improve efficiency in industries, households, and infrastructure.

Renewable Energy Integration: AI can facilitate the integration of renewable energy sources such as solar, wind, and hydroelectric power into the energy grid. AI algorithms can forecast energy generation, optimize grid management, and balance supply and demand to maximize the use of renewable energy while minimizing reliance on fossil fuels.

Circular Economy Practices: AI can support the transition to a circular economy by promoting reuse, recycling, and remanufacturing of products and materials. AI-powered systems can optimize waste management processes, identify opportunities for resource recovery, and facilitate the reuse of materials in the production cycle.

Climate Change Mitigation: AI can contribute to climate change mitigation efforts by analyzing climate data, modeling climate scenarios, and identifying strategies to reduce greenhouse gas emissions. AI algorithms can optimize transportation systems, improve energy efficiency, and support the development of low-carbon technologies and practices.

Sustainable Agriculture and Food Systems: AI can promote sustainable agriculture practices and food systems by optimizing crop management, reducing water usage, and minimizing chemical inputs. AI-powered precision agriculture technologies can analyze soil data, monitor crop health, and optimize irrigation and fertilization to improve yields while minimizing environmental impact.

Biodiversity Conservation: AI can support biodiversity conservation efforts by monitoring ecosystems, tracking wildlife populations, and identifying conservation priorities. AI algorithms can analyze satellite imagery, sensor data, and ecological models to assess habitat health, detect threats to biodiversity, and inform conservation strategies.

Resilient Infrastructure: AI can enhance the resilience of infrastructure against climate change impacts and natural disasters. AI-powered predictive analytics can assess infrastructure vulnerabilities, optimize maintenance schedules, and prioritize investments in climate-resilient infrastructure designs and upgrades.

Sustainable Urban Development: AI can support sustainable urban development by optimizing urban planning, transportation systems, and infrastructure investments. AI algorithms can analyze urban data, simulate development scenarios, and inform decisions to promote compact, walkable cities, efficient public transportation networks, and green infrastructure.

By prioritizing these AI goals, organizations, governments, and communities can harness the power of technology to meet their needs sustainably, ensuring a balance between economic growth, social equity, and environmental stewardship.

CHAPTER 5
SUSTAINABLE URBAN DEVELOPMENT

Artificial intelligence (AI) can play a pivotal role in promoting sustainable urban development by optimizing resource usage, enhancing efficiency, and improving quality of life for urban residents. Here are some AI goals for sustainable urban development:

Smart Infrastructure Management: AI can optimize the management of urban infrastructure such as water, energy, transportation, and waste systems. AI algorithms can analyze data from sensors, meters, and other sources to detect inefficiencies, predict maintenance needs, and optimize resource allocation, reducing costs and environmental impact.

Urban Mobility Optimization: AI can optimize urban transportation systems to reduce congestion, minimize emissions, and improve accessibility. AI-powered traffic management systems can analyze real-time traffic data, optimize traffic signals, and coordinate public transit services to optimize travel times and promote sustainable modes of transportation such as walking, cycling, and public transit.

Energy Efficiency and Renewable Integration: AI can improve energy efficiency in buildings and promote the integration of renewable energy sources into urban energy systems. AI algorithms can analyze building energy data, optimize HVAC systems, and recommend energy-saving measures to reduce energy consumption and carbon emissions. Additionally, AI can optimize the operation of renewable energy sources such as solar panels and wind turbines to maximize energy generation and grid stability.

Smart Urban Planning: AI can support data-driven urban planning decisions to create more sustainable and livable cities. AI algorithms can analyze urban data, demographic trends, and environmental factors to inform land use planning, zoning regulations, and infrastructure investments. AI-powered simulation models can also evaluate the potential impacts of different development scenarios on urban sustainability and resilience.

Resilient Infrastructure and Disaster Preparedness: AI can enhance the resilience of urban infrastructure against natural disasters and climate change impacts. AI-powered predictive analytics can assess infrastructure vulnerabilities, predict extreme weather events, and optimize emergency response plans to mitigate risks and ensure the continuity of essential services during emergencies.

Community Engagement and Inclusion: AI can facilitate community engagement and participation in urban development processes. AI-powered platforms can collect feedback from residents, identify community priorities, and incorporate citizen input into decision-making processes, promoting transparency, accountability, and social equity in urban development initiatives.

Environmental Monitoring and Pollution Reduction: AI can monitor environmental quality in urban areas and support efforts to reduce pollution and improve air and water quality. AI algorithms can analyze sensor data, satellite imagery, and environmental models to detect pollution hotspots, identify sources of contamination, and develop strategies to mitigate environmental risks and protect public health.

By prioritizing these AI goals, cities can harness the power of technology to address urban challenges and create more sustainable, resilient, and equitable communities for future generations.

CHAPTER 6
ENVIRONMENTAL MONITORING AND POLLUTION REDUCTION

Artificial intelligence (AI) can significantly contribute to environmental monitoring and pollution reduction efforts through data analysis, predictive modeling, and optimization of resource usage. Here are some AI goals for environmental monitoring and pollution reduction:

Real-Time Monitoring: Develop AI-powered systems for real-time monitoring of environmental parameters such as air quality, water quality, and soil contamination. These systems can integrate data from various sensors, satellites, and IoT devices to provide timely insights into pollution levels and environmental health.

Early Warning Systems: Implement AI algorithms to detect early signs of environmental degradation and potential pollution events. By analyzing historical data and environmental patterns, AI can forecast risks and issue early warnings to mitigate the impact of pollution incidents on ecosystems and public health.

Pollution Source Identification: Utilize AI techniques such as machine learning and pattern recognition to identify sources of pollution and trace pollutants back to their origin. AI algorithms can analyze data from multiple sources, including emissions data, weather patterns, and geographical information, to pinpoint sources of pollution and inform regulatory actions.

Optimized Pollution Control Strategies: Develop AI-based models to optimize pollution control strategies and interventions. AI algorithms can analyze pollution data, evaluate the effectiveness of different mitigation measures, and recommend cost-effective solutions to reduce pollution levels while minimizing adverse impacts on economic activities.

Smart Environmental Remediation: Implement AI-driven technologies for smart environmental remediation and cleanup efforts. AI-powered robots, drones, and autonomous vehicles can assist in the remediation of contaminated sites by mapping pollution hotspots, collecting samples, and deploying remediation techniques such as phytoremediation or bioremediation.

Predictive Modeling for Pollution Reduction: Develop predictive models using AI to forecast pollution levels and trends based on environmental factors, human activities, and climate change scenarios. These models can help policymakers and stakeholders anticipate future pollution risks and develop proactive measures to prevent pollution and protect ecosystems.

Public Awareness and Citizen Science: Leverage AI to engage the public in environmental monitoring and pollution reduction efforts. AI-powered citizen science platforms can empower individuals to collect and analyze environmental data, report pollution incidents, and participate in community-based monitoring initiatives, fostering environmental stewardship and collective action.

Regulatory Compliance Monitoring: Develop AI tools for regulatory agencies to monitor compliance with environmental regulations and enforce pollution control measures. AI algorithms can analyze large volumes of regulatory data, satellite imagery, and remote sensing data to identify non-compliance issues, prioritize enforcement actions, and ensure accountability in pollution control efforts.

By pursuing these AI goals, stakeholders can leverage technology to enhance environmental monitoring, reduce pollution, and promote sustainable management of natural resources for the benefit of present and future generations.

CHAPTER 7
AI-POWERED CITIZEN SCIENCE PLATFORMS

AI-powered citizen science platforms leverage the collective intelligence and participation of citizens to contribute to scientific research and environmental monitoring efforts. These platforms utilize artificial intelligence algorithms to analyze the data collected by volunteers, enabling scalable and efficient data processing. Here's how AI-powered citizen science platforms work:

Data Collection: Citizen scientists collect various types of data using sensors, mobile apps, or other tools provided by the platform. This data can include environmental observations such as air quality measurements, wildlife sightings, or water quality samples.

Data Submission: Once collected, citizen scientists submit their data to the platform, either manually or through automated data uploads. The platform aggregates and organizes the data for analysis.

Data Analysis: AI algorithms are employed to analyze the collected data at scale. These algorithms can perform tasks such as image recognition, pattern recognition, classification, and predictive modeling to extract valuable insights from the data.

Quality Control: AI-powered citizen science platforms often incorporate quality control mechanisms to ensure the accuracy and reliability of the collected data. This may include outlier detection, data validation checks, and crowd-sourced verification processes.

Scientific Research: The analyzed data is made accessible to researchers, policymakers, and the public through interactive dashboards, maps, or research databases. Scientists can use this data to study environmental trends, biodiversity patterns, or other research questions.

Feedback and Engagement: Citizen scientists receive feedback on their contributions and may engage in discussions or collaborations with researchers and other participants through the platform. This fosters a sense of community and encourages continued participation in citizen science initiatives.

Examples of AI-powered citizen science platforms include:

•**eBird**: A platform where birdwatchers worldwide can submit bird observations, which are then analyzed using AI algorithms to track bird populations and migration patterns.

•**iNaturalist**: A platform for documenting and identifying plants and animals through user-contributed observations. AI-powered image recognition helps identify species from user-submitted photos.

•**Citizen Weather Observer Program (CWOP)**: A network of weather enthusiasts who contribute weather observations from personal weather stations. AI algorithms analyze the data to improve weather forecasting models.

•**Plankton Portal**: A citizen science project where volunteers classify plankton images to help researchers study ocean ecosystems. AI algorithms assist in image classification and data analysis.

Overall, AI-powered citizen science platforms democratize scientific research, empower citizens to contribute to environmental monitoring efforts, and generate valuable data for scientific discovery and environmental conservation.

CHAPTER 8
STRATEGIES AND INCENTIVES TO REDUCE CONSUMPTION TO SUSTAINABLE LEVELS

Reducing consumption to sustainable levels requires a multifaceted approach that addresses both individual behavior and systemic factors. Here are several strategies and incentives to achieve this goal:

Education and Awareness: Develop educational campaigns to inform consumers about the environmental, social, and economic impacts of overconsumption. Raise awareness about sustainable lifestyles, resource conservation, and the benefits of mindful consumption.

Product Labeling and Transparency: Implement labeling systems that provide clear information about the environmental footprint of products, including their carbon emissions, water usage, and ethical sourcing practices. Encourage companies to disclose this information transparently to help consumers make informed purchasing decisions.

Price Signals: Adjust pricing mechanisms to reflect the true costs of production, including environmental externalities. This can involve implementing taxes on products with high environmental impacts (e.g., carbon taxes, plastic taxes) and subsidies for sustainable alternatives (e.g., renewable energy subsidies, eco-friendly product subsidies).

Incentives for Sustainable Behavior: Provide incentives for consumers to adopt sustainable behaviors, such as offering discounts or rewards for recycling, using public transportation, or purchasing eco-friendly products. Reward programs, cashback incentives, and loyalty points can encourage consumers to make more sustainable choices.

Regulatory Measures: Enact regulations and policies that promote sustainable consumption practices, such as banning single-use plastics, implementing mandatory recycling programs, or setting targets for resource efficiency. Ensure enforcement mechanisms are in place to monitor compliance and penalize non-compliance.

Circular Economy Initiatives: Encourage the transition to a circular economy by promoting product reuse, repair, and recycling. Support initiatives such as product take-back programs, extended producer responsibility (EPR) schemes, and circular design principles to minimize waste and maximize resource efficiency.

Sharing Economy Platforms: Facilitate the sharing of resources and assets through sharing economy platforms, such as car-sharing services, tool libraries, or clothing rental platforms. Promote collaborative consumption to reduce the need for ownership and minimize resource consumption.

Community-Based Initiatives: Foster community engagement and grassroots initiatives that promote sustainable consumption practices at the local level. Support community gardens, food cooperatives, and repair cafes to encourage resource sharing, local production, and community resilience.

Green Procurement Policies: Implement green procurement policies for governments, businesses, and institutions to prioritize the purchase of sustainable and environmentally friendly products. Set sustainability criteria for procurement contracts and prefer suppliers with strong environmental and social credentials.

Corporate Responsibility: Encourage businesses to adopt sustainable practices throughout their supply chains, including reducing waste, improving energy efficiency, and sourcing ethically and sustainably produced materials. Recognize and reward companies that demonstrate leadership in sustainability.

By combining these strategies and incentives, policymakers, businesses, and individuals can work together to reduce consumption to sustainable levels and transition towards a more equitable and environmentally sustainable society.

CHAPTER 9
MAKING SUSTAINABLE HOUSING

Creating sustainable housing involves setting clear goals that prioritize energy efficiency, resource conservation, environmental responsibility, and social equity. Here are some goals for making sustainable housing:

Energy Efficiency: Set goals for energy-efficient design and construction to minimize energy consumption and reduce greenhouse gas emissions. This includes achieving high levels of insulation, using energy-efficient appliances and lighting, and incorporating renewable energy sources such as solar panels.

Resource Conservation: Establish goals for resource-efficient building practices to minimize waste and maximize the use of sustainable materials. This may involve using recycled or locally sourced materials, optimizing construction processes to reduce material waste, and incorporating water-saving fixtures and systems.

Environmental Responsibility: Set goals to minimize the environmental impact of housing developments, including reducing carbon emissions, preserving natural habitats, and mitigating the urban heat island effect. This may involve incorporating green spaces, planting native vegetation, and implementing stormwater management strategies.

Affordability and Accessibility: Ensure that sustainable housing is accessible and affordable to people of all income levels. Set goals for providing affordable housing options with low operating costs, incorporating universal design principles to accommodate people with disabilities, and promoting equitable access to public transportation and amenities.

Resilience to Climate Change: Establish goals for building resilience to climate change impacts, including extreme weather events, rising temperatures, and sea-level rise. This may involve designing buildings to withstand extreme weather conditions, incorporating passive cooling and heating strategies, and locating developments away from flood-prone areas.

Community Engagement and Social Equity: Set goals for fostering community engagement and promoting social equity within housing developments. This may involve involving residents in the design and planning process, creating shared spaces and amenities that encourage social interaction, and ensuring that housing developments are inclusive and accessible to people from diverse backgrounds.

Long-Term Durability and Maintenance: Establish goals for ensuring the long-term durability and maintenance of sustainable housing developments. This includes using durable materials, designing buildings for easy maintenance and repair, and providing residents with resources and support for ongoing maintenance and upkeep.

Certification and Standards Compliance: Set goals for achieving certification or compliance with sustainable building standards and certifications, such as LEED (Leadership in Energy and Environmental Design), Passive House, or the Living Building Challenge. This can help ensure that housing developments meet high standards of sustainability and environmental performance.

By setting clear goals for sustainable housing and incorporating principles of energy efficiency, resource conservation, environmental responsibility, and social equity, stakeholders can work towards creating housing that meets the needs of both current and future generations while minimizing environmental impact.

CHAPTER 10
EDUCATING TO MEET NEEDS SUSTAINABLY

Educating individuals and communities to meet their needs sustainably is crucial for building a more environmentally conscious and resilient society. Here are some goals for sustainable education:

Raise Awareness: Increase awareness about the importance of sustainability and its relevance to everyday life. Goal: Ensure that individuals understand the connections between their actions and their impact on the environment, society, and economy.

Promote Critical Thinking: Foster critical thinking skills to evaluate information, analyze complex issues, and make informed decisions. Goal: Empower individuals to question conventional practices, explore alternative solutions, and consider the long-term consequences of their actions.

Encourage Systems Thinking: Introduce systems thinking concepts to understand the interconnectedness of environmental, social, and economic systems. Goal: Help individuals recognize the interdependencies between different aspects of sustainability and identify leverage points for positive change.

Provide Practical Skills: Offer practical skills and knowledge to implement sustainable practices in daily life. Goal: Equip individuals with skills such as composting, gardening, energy conservation, waste reduction, and sustainable transportation to reduce their ecological footprint and promote sustainable living.

Promote Environmental Literacy: Foster environmental literacy by teaching about ecosystems, biodiversity, climate change, and other environmental issues. Goal: Ensure that individuals have a basic understanding of environmental concepts and are informed about current environmental challenges and opportunities.

Cultivate Empathy and Values: Cultivate empathy, compassion, and a sense of responsibility towards the environment and future generations. Goal: Foster values such as stewardship, respect for nature, and a commitment to social and environmental justice.

Encourage Active Citizenship: Encourage active citizenship and civic engagement in sustainability initiatives and community projects. Goal: Empower individuals to participate in local decision-making processes, advocate for sustainable policies, and contribute to collective efforts for positive change.

Integrate Sustainability Across Disciplines: Integrate sustainability principles into education across disciplines, including science, technology, engineering, mathematics (STEM), social sciences, humanities, and arts. Goal: Ensure that sustainability education is interdisciplinary and reflects the complexity of real-world sustainability challenges.

Promote Lifelong Learning: Foster a culture of lifelong learning and continuous improvement in sustainability knowledge and practices. Goal: Encourage individuals to seek out opportunities for ongoing education, skill development, and engagement in sustainability-related activities throughout their lives.

Lead by Example: Lead by example and create a culture of sustainability within educational institutions, workplaces, and communities. Goal: Demonstrate sustainable practices, provide leadership opportunities, and create supportive environments that enable individuals to adopt sustainable behaviors and inspire others to do the same.

By setting and working towards these goals, educators, policymakers, and community leaders can empower individuals to meet their needs sustainably, contribute to positive societal change, and build a more resilient and equitable future for all.

CHAPTER 11
MEDIA TO MEET NEEDS SUSTAINABLY

To meet needs sustainably and promote a transition away from fossil fuel media, it's essential to support and develop alternative media sources that prioritize sustainability, environmental responsibility, and social justice. Here are some media initiatives and platforms that can help achieve these goals:

Independent and Nonprofit Media Outlets: Support independent and nonprofit media organizations that prioritize investigative journalism, environmental reporting, and coverage of sustainability issues. These outlets often provide in-depth analysis and critical perspectives on environmental and social issues, without the influence of corporate interests.

Environmental and Climate-Focused Publications: Promote media outlets and publications that specialize in environmental and climate reporting, such as magazines, websites, and podcasts dedicated to sustainability, conservation, and climate change. These platforms can raise awareness about environmental challenges, highlight solutions, and empower readers to take action.

Community-Based Media: Support community-based media initiatives that amplify the voices of marginalized communities and promote local solutions to environmental and social issues. Community radio stations, newspapers, and online platforms can provide a platform for grassroots activism, citizen journalism, and community engagement on sustainability issues.

Digital Platforms for Environmental Content: Explore digital platforms and online communities that curate and share environmental content, such as documentaries, podcasts, and webinars on sustainability topics. Platforms like YouTube, Netflix, and TED Talks feature a wide range of content related to sustainability, renewable energy, conservation, and environmental activism.

Social Media and Digital Activism: Utilize social media platforms and digital activism tools to raise awareness about sustainability issues, advocate for policy change, and mobilize grassroots action. Social media campaigns, hashtags, and online petitions can amplify environmental messages, engage diverse audiences, and catalyze social change.

Green Tech and Innovation Media: Support media coverage of green technology, innovation, and renewable energy solutions. Platforms that focus on clean energy, sustainable transportation, circular economy initiatives, and eco-friendly innovations can inspire positive change and highlight opportunities for a sustainable future.

Environmental Documentaries and Films: Promote environmental documentaries, films, and series that raise awareness about environmental issues, inspire action, and drive change. Documentaries such as "An Inconvenient Truth," "Before the Flood," and "Our Planet" have played a critical role in shaping public discourse on climate change and environmental conservation.

Educational Resources for Sustainability: Explore educational resources and materials on sustainability topics, such as online courses, educational websites, and digital libraries. Platforms like Coursera, edX, and Khan Academy offer courses and resources on sustainability, environmental science, renewable energy, and sustainable development.

By supporting and promoting these alternative media sources and initiatives, individuals can contribute to a more informed, engaged, and sustainable society, while reducing reliance on fossil fuel-dependent media outlets. Additionally, advocating for policies that support media diversity, transparency, and accountability can help foster a media landscape that prioritizes sustainability and social responsibility.

CHAPTER 12
WHAT FACTORS WILL LIMIT AI?

Artificial Intelligence (AI) is a rapidly evolving field, but there are several factors that can limit its development and deployment:

Data Quality and Quantity: AI models require large amounts of high-quality data to learn effectively. Limited or biased data can lead to inaccurate or unfair outcomes.

Algorithmic Limitations: The effectiveness of AI algorithms can be limited by their design and complexity. Some tasks may require more sophisticated algorithms that are computationally expensive or difficult to develop.

Computational Resources: Training complex AI models requires significant computational resources, including powerful processors and large amounts of memory. Limited access to such resources can hinder the development of AI systems.

Ethical and Social Implications: AI systems can raise ethical concerns related to privacy, fairness, and accountability. Addressing these concerns may require regulatory frameworks and societal consensus, which can slow down the adoption of AI technologies.

Interpretability and Transparency: AI models often operate as "black boxes," making it difficult to understand how they arrive at their decisions. Lack of interpretability can limit trust and acceptance of AI systems, especially in critical domains like healthcare and criminal justice.

Robustness and Security: AI systems are vulnerable to adversarial attacks, where malicious actors manipulate input data to deceive the model. Ensuring the robustness and security of AI systems is crucial for their safe deployment in real-world applications.

Human-AI Collaboration: Integrating AI systems into existing workflows and processes can be challenging, requiring changes in organizational culture and skill sets. Effective human-AI collaboration is essential for maximizing the benefits of AI technology.

Regulatory and Legal Constraints: Regulatory frameworks may impose restrictions on the development and deployment of AI systems, particularly in sensitive areas such as healthcare and finance. Compliance with these regulations can pose challenges for AI developers and users.

Resource Allocation and Funding: Adequate funding and support are necessary for advancing AI research and development. Limited resources may constrain progress in developing more advanced AI technologies.

Cultural and Societal Factors: Societal attitudes and cultural norms can influence the adoption and acceptance of AI technologies. Addressing cultural barriers and fostering public trust is essential for the widespread adoption of AI solutions.

Overall, addressing these factors requires a multidisciplinary approach involving collaboration between researchers, policymakers, industry stakeholders, and the broader community to realize the full potential of AI while mitigating its limitations and risks.

CHAPTER 13
WHAT PRODUCTION AND DELIVERY PROBLEMS COULD DISRUPT AI?

Several production and delivery problems could disrupt the advancement and deployment of AI technologies:

Supply Chain Disruptions: AI hardware components, such as processors and GPUs, rely on complex global supply chains. Disruptions in the supply chain, whether due to natural disasters, geopolitical tensions, or other factors, can lead to shortages and delays in manufacturing AI systems.

Quality Control Issues: Ensuring the quality and reliability of AI hardware and software components is crucial for their effective deployment. Quality control issues, such as defects in hardware or bugs in software, can lead to product recalls, delays, and increased costs.

Data Availability and Quality: AI models depend on large amounts of data to learn and make accurate predictions. Issues related to data availability, such as data silos or inadequate data labeling, can hinder the development and deployment of AI systems.

Data Privacy and Security Concerns: Data privacy regulations, such as the GDPR in Europe, impose strict requirements on the collection, storage, and processing of personal data. Ensuring compliance with these regulations while maintaining data security can pose challenges for AI developers and users.

Ethical and Bias Concerns: AI systems can exhibit biases and unfairness, particularly when trained on biased data or developed without considering ethical implications. Addressing ethical concerns and mitigating bias in AI systems is essential for ensuring their acceptance and trustworthiness.

Regulatory Compliance: Regulatory frameworks governing AI technologies vary across jurisdictions and industries. Ensuring compliance with applicable regulations, such as safety standards and certification requirements, is essential for the lawful deployment of AI systems.

Skills Shortage: Building and deploying AI systems require specialized skills in areas such as machine learning, data science, and software engineering. A shortage of skilled professionals in these fields can limit the scalability and adoption of AI technologies.

Infrastructure Limitations: AI systems often require robust infrastructure for training and inference, including high-performance computing resources and scalable storage solutions. Inadequate infrastructure can limit the scalability and performance of AI deployments.

Interoperability Challenges: Integrating AI systems with existing IT infrastructure and workflows can be complex, particularly in heterogeneous environments. Interoperability challenges can arise from differences in data formats, protocols, and system architectures.

Cost Constraints: AI technologies can be costly to develop, deploy, and maintain, particularly for small and medium-sized enterprises with limited resources. Cost constraints can limit the adoption of AI technologies, especially in resource-constrained environments.

Addressing these production and delivery problems requires collaboration between AI developers, policymakers, industry stakeholders, and the broader community to develop scalable, reliable, and ethical AI solution.

CHAPTER 14
OPERATIONAL LEARNING

Operational learning is a concept that refers to the process of continuously improving and optimizing operations through real-time feedback and adaptation. In the context of artificial intelligence and machine learning, operational learning involves deploying AI systems in real-world environments where they can gather data, learn from experience, and dynamically adjust their behavior to improve performance over time.

Key characteristics of operational learning include:

Real-Time Feedback: Operational learning systems receive feedback from their environment or users in real time, allowing them to quickly adapt to changing conditions and optimize their performance accordingly.

Adaptation and Evolution: Operational learning systems can evolve and improve over time by continuously updating their models and algorithms based on new data and experiences.

Iterative Improvement: Operational learning involves an iterative process of experimentation, learning, and refinement, where AI systems gradually improve their performance through repeated cycles of feedback and adaptation.

Deployment in Production Environments: Operational learning occurs in real-world production environments, rather than controlled laboratory settings, allowing AI systems to learn from the complexities and uncertainties of real-world data and interactions.

Scalability and Efficiency: Operational learning systems are designed to scale efficiently to handle large volumes of data and interactions, enabling them to learn and adapt in real time across diverse use cases and scenarios.

Operational learning has applications across various domains, including manufacturing, logistics, healthcare, finance, and cybersecurity. By enabling AI systems to continuously learn and adapt in real-world environments, operational learning can drive innovation, efficiency, and competitive advantage for organizations seeking to leverage AI technologies to optimize their operations.

CHAPTER 15
WHAT IS MACHINE LEARNING?

Machine learning is a subset of artificial intelligence (AI) that focuses on the development of algorithms and statistical models that enable computers to learn from and make predictions or decisions based on data, without being explicitly programmed to perform specific tasks. In essence, machine learning algorithms use data to learn patterns, relationships, and insights, which they can then use to make predictions or decisions.

Key components of machine learning include:

Data: Machine learning algorithms require data to learn from. This data can include structured data (e.g., numerical values in a database) or unstructured data (e.g., text, images, audio).

Features: Features are the variables or attributes within the data that the machine learning algorithm uses to make predictions or decisions. Feature selection and engineering are crucial steps in preparing data for machine learning models.

Model Training: During the training phase, machine learning algorithms are exposed to labeled data (supervised learning) or unlabeled data (unsupervised learning). The algorithm learns from this data to identify patterns and relationships that can be used to make predictions or extract insights.

Model Evaluation: After training, machine learning models are evaluated on their performance using separate datasets (validation and test sets). Performance metrics such as accuracy, precision, recall, and F1 score are used to assess the model's effectiveness.

Model Deployment: Once trained and evaluated, machine learning models can be deployed to make predictions or decisions in real-world applications. Deployment involves integrating the model into existing systems or workflows and ensuring its scalability, reliability, and security.

Machine learning techniques can be broadly categorized into three main types:

Supervised Learning: In supervised learning, the algorithm learns from labeled data, where each data point is associated with a target label or outcome. The goal is to learn a mapping from input features to output labels, allowing the algorithm to make predictions on new, unseen data.

Unsupervised Learning: In unsupervised learning, the algorithm learns from unlabeled data, identifying patterns, clusters, or structures within the data. Unsupervised learning techniques include clustering, dimensionality reduction, and anomaly detection.

Reinforcement Learning: In reinforcement learning, the algorithm learns through interaction with an environment, receiving feedback in the form of rewards or penalties for its actions. The goal is to learn a policy that maximizes cumulative reward over time, enabling the algorithm to make sequential decisions in dynamic environments.

Machine learning has applications across various domains, including healthcare, finance, e-commerce, cybersecurity, autonomous vehicles, natural language processing, and computer vision. It is a powerful tool for extracting insights from data, automating tasks, and making data-driven decisions in complex and uncertain environments.

CHAPTER 16
WHAT IS THE POLITICAL ECONOMY OF AI?

The political economy of artificial intelligence (AI) refers to the intersection of politics, economics, and technology in the development, deployment, and governance of AI systems. It involves examining how power dynamics, economic interests, and policy decisions shape the development and impact of AI technologies on society.

Key aspects of the political economy of AI include:

Power Dynamics: AI technologies can influence power dynamics within and between nations, organizations, and social groups. The ability to develop and deploy AI systems can confer strategic advantages in areas such as national security, economic competitiveness, and social control.

Economic Implications: AI has significant economic implications, affecting industries, labor markets, and global trade. AI technologies have the potential to increase productivity, create new job opportunities, and drive economic growth, but they also raise concerns about job displacement, income inequality, and concentration of wealth.

Regulatory Environment: The regulatory environment for AI varies across countries and regions, reflecting different approaches to balancing innovation with concerns about safety, fairness, and accountability. Regulatory frameworks may address issues such as data privacy, algorithmic bias, cybersecurity, and ethical use of AI technologies.

Global Competition: AI is a highly competitive field, with countries and companies vying for leadership in research, development, and market dominance. Geopolitical tensions and economic rivalries can influence investment priorities, intellectual property rights, and international collaborations in AI.

Labor and Workforce: The widespread adoption of AI technologies has implications for the future of work and the workforce. While AI can automate routine tasks and increase efficiency, it can also lead to job displacement and changes in skill requirements. Policies related to education, training, and social safety nets are crucial for supporting workers in the transition to an AI-driven economy.

Ethical and Social Impacts: AI raises ethical and social concerns related to privacy, fairness, transparency, accountability, and autonomy. Political debates and public discourse play a crucial role in shaping norms, values, and regulations governing the responsible development and use of AI technologies.

International Cooperation: Given the global nature of AI research, development, and deployment, international cooperation is essential for addressing common challenges and ensuring that AI benefits are shared equitably across countries and regions. Collaboration on standards, best practices, and knowledge sharing can promote responsible AI governance at the global level.

Understanding the political economy of AI requires analyzing the interplay of economic interests, policy choices, societal values, and technological capabilities in shaping the trajectory of AI development and its impact on individuals, communities, and nations. It involves navigating complex trade-offs between innovation, regulation, and social welfare to harness the potential of AI for the benefit of all.

CHAPTER 17
CLASS PROBLEMS

AI has the potential to exacerbate existing class problems or socioeconomic inequalities in several ways:

Labor Market Displacement: AI technologies can automate routine tasks and jobs, leading to job displacement and unemployment, particularly for low-skilled workers in sectors such as manufacturing, retail, and transportation. Workers without access to education or training programs to develop new skills may be left behind, widening the gap between those with and without employable skills.

Concentration of Wealth: AI has the potential to concentrate wealth and power in the hands of a few tech companies and wealthy individuals who control AI technologies and platforms. This concentration of economic power can exacerbate income inequality and further marginalize disadvantaged communities who lack access to AI-driven opportunities for economic advancement.

Access to AI Education and Training: Access to education and training in AI-related fields is often limited to individuals from privileged socioeconomic backgrounds who can afford expensive tuition fees, advanced technology, and access to high-quality educational resources. This lack of access to AI education and training opportunities can perpetuate class divides and hinder social mobility for marginalized communities.

Algorithmic Bias and Discrimination: AI systems can perpetuate and even amplify biases present in training data, leading to discriminatory outcomes in areas such as hiring, lending, criminal justice, and healthcare. Biased AI algorithms can disproportionately harm marginalized communities, exacerbating existing inequalities and reinforcing systemic injustices.

Surveillance and Control: AI-powered surveillance technologies, such as facial recognition and predictive policing systems, can disproportionately target and surveil marginalized communities, leading to increased social control and oppression. These technologies can amplify existing power imbalances and infringe on individuals' privacy rights, particularly those from disadvantaged socioeconomic backgrounds.

Access to AI-Driven Services: AI-driven services and platforms, such as personalized healthcare, financial advice, and educational resources, may be inaccessible to individuals without adequate digital literacy, internet access, or financial resources. This digital divide can further marginalize underserved communities and perpetuate disparities in access to essential services and opportunities.

Addressing the class problems exacerbated by AI requires proactive measures to mitigate risks and promote equitable access to AI-driven opportunities. This includes implementing policies to regulate AI technologies, ensure transparency and accountability in algorithmic decision-making, invest in education and training programs for underserved communities, and promote diversity and inclusion in the AI workforce. Additionally, fostering interdisciplinary collaboration and community engagement is crucial for developing AI solutions that address the needs and priorities of all members of society.

CHAPTER 18
HOW TO END ECONOMIC CLASS DIFFERENCES IN THE US AND AROUND THE WORLD?

AI technology can help identify and mitigate biases in data sets, algorithms, and decision-making processes. By implementing fairness and transparency measures in AI systems, organizations can work towards reducing discriminatory outcomes. Additionally, AI can be used to promote diversity and inclusion by analyzing patterns of bias and providing insights for creating more equitable policies and practices. It is important to note that while AI can be a tool in combating bigotry, it is essential for human oversight and ethical considerations to be integrated into the development and deployment of AI systems.

I appreciate your understanding of how AI can be utilized to combat bigotry. It is crucial for organizations to prioritize fairness, transparency, and diversity when implementing AI solutions to address biases effectively. By incorporating ethical considerations and human oversight, we can ensure that AI technologies are used responsibly and contribute to creating a more inclusive and equitable society. Thank you for your interest in this important topic.

Addressing economic class differences in the US and globally is a complex and multifaceted issue that requires a comprehensive approach. Some strategies that can be considered include implementing policies that promote income equality, such as progressive taxation and minimum wage laws. Investing in education and job training programs to provide equal opportunities for all individuals regardless of their socioeconomic background can also help reduce economic disparities. Additionally, fostering entrepreneurship, supporting small businesses, and promoting financial literacy can empower individuals to improve their economic status. Collaboration between governments, businesses, and civil society is essential to create sustainable solutions that address economic class differences effectively.

Economic incentives to reduce and stop economic social and political bias toward women

One approach to reducing economic, social, and political bias towards women is to implement economic incentives that promote gender equality. This can include policies such as equal pay for equal work, providing financial incentives for companies that demonstrate gender diversity in leadership positions, and offering tax breaks or subsidies for businesses that prioritize gender equality in their hiring practices. Additionally, investing in programs that support women entrepreneurs, providing access to affordable childcare and parental leave policies, and offering training and education opportunities for women in traditionally male-dominated industries can help address bias and promote gender equality in economic, social, and political spheres. It is essential to combine these economic incentives with broader societal efforts to challenge stereotypes, promote diversity, and create a more inclusive environment for women to thrive.

CHAPTER 19
AI ACCELERATES CAPITALISM AND GLOBAL WARMING

AI is a catalyst for capitalism. It makes production more efficient by using computer-controlled machines to make more things of another quality with less time and money. Under appropriate leadership, AI can advance sustainable economic policy that prioritizes using AI to create an economy that meets needs without destroying our means to live. Without government leadership to curb the race to an AI gold mine, millions, perhaps billions, will face nonstop disruption to a decent life. Quickly, the AI machines are conducting humans day and night as people from all walks of life scramble and race to get their hands on the Internet code to get hooked in as global warming pummels people out of existence.

What is the AI goldmine?

About 50 years ago, union workers who manufactured cars could make a living to support a family. Gradually, as the industry replaced workers with robots, most of those jobs disappeared as corporations cashed in the robot gold mines that don't have an hourly wage, health insurance, or a pension plan. The gold mine is the illusive profit, the jackpot in the sky that they clutch for to secure the endless flow of money as they produce a higher quality car for less money but can still charge the price for making a car with human labor. So even though they can create a better car for less money, the consumer doesn't get a share of the fruits of the robots.

What does AI have to do with global warming?

Global warming increases the frequency and intensity of storms and destabilizes the climate. It destroys the environmental conditions for people to live. AI makes capitalism go faster. Machines produce better products faster and cheaper than people. AI can accelerate the transition to a sustainable economy. Governments can set targets to power vehicles and buildings without using fossil fuels. With incentives, the process could be accelerated. As this transition unfolds, the government could limit non-sustainable economic production until a sustainable economy has been set up.

Like a seduction to a trap, humans get sucked into AI.

I'm in it. I saw the advantages and used them. Now, I don't want to stop. No one told me I had a choice. Why would I make my life more difficult? I used to flip out when I had to go somewhere new. I would get anxious. What if I got lost and drove my car around, wasting fossil fuels, increasing pollution, and being late for my appointment? What a drag that is. When you have a navigator or a computer, it's perfect; nothing can go wrong. Like so many things that look rosy in the beginning and show another side as time passes. The computer knows where to go. I type in the address, and bam, I've got directions! That's the myth. Especially when technology is new, there are bugs in it. You need to know what they are and when they will appear, and you need to know the problem. Is it me, the data connection, the software, or how to use it? A lot of it is about the interface. There's another world of zeros and ones where engineers have figured out how to process the data and spit out a solution. That's done by people who set up the algorithms for the machine to learn how to do it. The technology may contain peculiarities of the people who made it. The machine can account for this. If the user says it worked, maybe it did. Or it works for some people and not others.

Without data, AI does nothing; with data, it does something.

There are hooks everywhere enticing you to use it. You are the fish it's looking for. Humans provide the data, which is the gold of the AI economy, but the data is piling up, and now it's a mountain more significant than a mountain from here to the moon. So, what do we do with the data? How can the data be used? It's a vast mass, and it's getting out of control. Dirt was only valuable once people learned how to use it. In permaculture, the problem is the solution. How can all that useless data be used? Some use it to predict the future. AI uses the data to make predictions. By surveying large amounts of data, patterns can be noticed. The patterns are used to predict the future.

The information is free; searches and data come out quickly, efficiently, and easily accessible.

Indeed, I still need to do something to get the information. I set up my computer, searched, and KaBlamo! I got the information in the blink of a second. Years ago, that same task could take me years, but now it can be done quickly. Indeed, a more enticing seduction could rarely, if ever, be found. Once I have my computer set up, I collect the gold day and night. I get the information and store it. It's all free. I can have as much as I want and don't pay anything for it. That's a seduction. Infinite whatever I want. That's the illusion and seduction. I have the pot of gold, the one that never stops spewing out the gold.

The seduction

If it makes you click or makes you stay, it's working. People are following something on the screen. It's working to collect data to predict how you will behave. Will you buy now or later? It's a 0 and a 1. Anything that can be made into this or that, one thing or the other, can be turned into a click, which is money. It may be .ooo1 cents per click, and it adds up, especially if you have 3 billion people doing it all day without getting paid. Keep them clicking and Ka-ching; it will happen. The bank accounts will swell, and people will keep going, keep buying, and keep the rat race going.

The addiction

The goal is to mine attention. They turn attention into money. It knows more about you than you do. Gold used to be oil, but now the gold is in the brain. Generally, if they can get one click out of you, you are finished. All exists in the mind first and elsewhere later. This is why the click is money. Click money is like a neural connection with the potential to turn it into cash. Engineers, psychologists, writers, and a whole army of people in the corporation work day and night on this.

Do you think you have the advantage or that you are in control? Perhaps you have some power, and maybe you have no chance of negotiating. Ultimately, you come back to AI. When it tells you to go, you go, or you don't get what you want. Perhaps you want to fix your computer or find out how to help someone who can't breathe. It's life-saving and great, but you have to do the click.

Click Money

How can you turn a click into money? Is it one click or infinite clicks? One and infinite are the same. It's pushing down; it's a neural link between something in the brain and something on the screen. It's a mediation to dissolve the person into what's behind the screen. People die, or their computers break, and they can't or don't click anymore. They get clicked out.

Resisting the click

How does the click become money? How can I type keywords into a search and get that turned into money? How can that happen? What is the magic there? I'm searching for the definition of the word "the," and bam, it's right there; then I turn it off. No harm was done anywhere. I'm still in control. I didn't buy anything; I don't have a credit card. I don't own a computer; I use the one at the library. There's no way I can buy anything. So what's the problem? Some people don't buy anything and aren't plugged into the system. It happens, but it is very uncommon. You may need something like a pair of shoes. So you look around at all the shoe stores, and they are pretty good, and you try them on and buy them at the store. Shoes are an excellent thing to buy at a store. Also, I don't buy anything. I'm committed to resisting the system, so what's the problem?

Can I resist forever?

You don't use or buy anything with a computer. You are resisting the screen thing and are good at it; you will make it happen, and it will work—end of story. You live for a while, and you die and are put in your grave, and that's it. I'm not a screenhead; I have never been and will never be one. Can that work? It does work for some people. For most, this is nearly impossible or undesirable.

Screen Heads and meditation

If you are forever focused on a screen, will you become a screenhead? Can your mind become a screen or that thing behind the screen that you chase day and night? Is any thought possible without being filtered by a screen? When I think of AI, all of what comes to mind comes through a computer, but what if I wake up and the first thing I do is to meditate on what comes to mind and I write it down on paper without a screen and then I avoid the screen as long as possible. I use the initial daily meditation to control the screen. This must be a reversal; after all, the only thing I have to do to be in control is to shut it off or destroy it. I could give away all my computers, phones, and screens, and then what will happen? It's possible to do something like that, and some people do it. In this society, it's becoming more and more challenging to do something like that.

Mediation

Are they opposites? Mediation is what's in between. It's in-between that something in your mind and something else out there. All is mediation. It's only possible to think of something with an in-between. What's the in-between? It's everything between the thought and something out there: mediation. I'm thinking about the sunshine. So what's between me and the sunshine? Nothing, of course; it's just me and the sunshine, a simple, immediate, and direct connection. There's a song, maybe a scene from a movie where the hot sun finally comes after a forever winter. There's a memory of the never-ending heat, the hot sun coming through the window on a fantastic day. The burning hot sun of global warming is slowly erasing what stability we force onto what is out there. Is AI the mediation between me and what's out there? I'm typing and not connected to the internet. AI can't get me now, but what about when I look at the news? AI is between me and the news. It's this in-between machine. It's collecting my clicks, counting the seconds, counting how long the machine can keep me occupied. It occupies my attention; it's a takeover. It's grabbing me like an alligator, like quicksand, bam. It's got me. I'm in.

Is it using me, or am I using it?

Who's got the power? Me or AI. I'll pull the plug, and bam, AI is finished! The answer goes both ways. If you want directions to visit a friend, you find a program on the internet, type both addresses, and quickly, in the blink of an eye, you get results. You just got some beneficial information very quickly. People who have never lived without a mapping program on the internet have no idea what a chore it could be to find a destination. That information can save time and money, depending on your actions. Before computers gave driving information, you had to buy a map and determine the best way to reach your destination. It was tedious work. The map may not be up to date or show all the information you need, or if you can't see, then what? It could be a struggle, and you could end up lost and trying to find where to go. With patience and persistence, you could make it to your destination. It would be more difficult if you were driving alone, especially if there's a lot of traffic. In this case, I've given information to AI. It would know where I'm coming from and going to, especially if I have set up my phone to be tracked. Am I giving up a lot of information? It depends on what you are doing. If someone else can find and use that information, and I don't want them to know, I am giving up a lot of information. Let's say that I'm in an abusive relationship, and the person abusing me wants to track me, and somehow, they know how to get the information. This could cause significant problems for me. Would the AI be using me? Or would I be using AI? I'm using AI to get the driving instructions. I give the information to the computer. The computer has not asked for the information, and even if it did, I'm still giving it to the computer, but am I? In this case, I may not know that my abuser has access to the information. I think I'm choosing to use a free computer service, but I don't know that I'm choosing to give the information to the abuser. I don't know how the computer will use the information. There are a lot of things the computer could do with the information. It could sell it to the highest bidder. It could give it to people who could use it against me, even though I've done nothing wrong. It may not be possible to track what happens with the information. They could be stored in a device that's encrypted and impossible for me to access. Again, the only person who could access it may be the highest bidder or law enforcement. What if law enforcement uses it by mistake, and now they think I'm a dangerous, wanted criminal? While that may be very unlikely, it does happen to people who have to pay a very high price for wanting to use some "free Information."

Do I have a choice?

Can you choose not to walk on the sidewalk, live in a house, or use the benefits of this society? Yes, you can, but it may make your life more difficult or tedious or be the perfect life for you. If you have to make some tea, how will you do it?

1. Put herbs in water, wait 10 hours, and then drink it.
2. Build a fire outside, heat water, and drink tea.
3. Use a gas or electric stove or microwave to heat water, make tea, and drink it.
4. Use an automatic tea maker, press a button, and it's done.

The list uses the least inputs to the most inputs on the bottom. Cold brewing is the first method where no fire is used; it can be done anywhere, it needs water and herbs, and there is no outside energy source. It's free to find the tea leaves outside or use whatever plants you have to make a beverage. This method takes the most time, but it is the least amount of work overall. The second method, which is to build a fire outside and heat the water, seems the most primitive, uses the most labor, time, and materials, and generates the most pollution. Method one requires the least number of other people and machines. Method 2 is similar but requires a way to start the fire. Method number 3 is in another category. Here, an elaborate system is set up to produce and deliver the energy required to heat the water. Mining natural gas requires many machines and systems. After the energy is made, the gas and electric lines fuel the equipment. This method requires setting up a society that can easily break down and requires extensive time, labor, and materials. It's a demanding system with many elements in a long chain of production that can break down. The last method uses the least amount of human labor to make the tea and the most amount of human time, energy, and money to set it up. First, everything in method three must be set up in advance. Electricity is typically used to power automated systems. Then, the automated coffee maker must be produced. It's another level of sophistication after making a standard coffee maker without electronic features. Making the most automated method of making teas requires the most setup in terms of human time, energy, labor, and materials. It's also the method that is most likely to break down. The more elements and moving parts, the more likely it will break down.

Is it only about convenience?

One difference between what artificial intelligence offers and what it is connected to is convenience. Using a clothes washer is easier than doing it by hand. It's easier to use a mapping program than to ask people, find a map and hope it's up to date, or get a big idea and, by trial and error, find the place I'm going to. Relax and enjoy the ride; we will still get there without a mapping program. Or is it life and death? If I find a surgeon or herbalist with precious skills and experience, my time will be well-spent. I've been avoiding learning rudimentary programming, which could save me time. Instead, I will do the heavy work for a while and do what I could do automatically by hand and enjoy those moments of easy work where I tidy things up and make them look more enticing.

CHAPTER 20
WHAT IS AI?

What is artificial?

Artificial means what is produced by humans instead of what nature makes. A tree is natural; a cabinet is artificial. The human brain is natural; a computer is artificial. Humans are intelligent; computers contain artificial intelligence. When a machine does it, it's artificial; when a human does it, it's intelligent. The distinctions seem arbitrary. Humans are nature. What humans produce is nature. What humans produce does not exist on other planets. We don't know. It's easy to assume that humans do not exist on another planet. Whatever people produce is human. It cannot exist without humans; therefore, it is human.

What is intelligence

The ability to acquire and apply knowledge and skills is intelligence. We watch other people do something, and then we do it. For some, it's a quick way to learn. I grew up doing it. Now it's comfortable and easy. Learning from a manual is another story. We learn something basic and straightforward. We use that to learn a bit more. We find out we need to get it right, so we make adjustments and see if that gets closer to the goal. We get stuck again. No solution presents itself. Over and over, the process happens. With experience and repeating the process, we make solutions as the problems change.

What is artificial intelligence?

Artificial intelligence uses computers, robots, and drones to perform human activities. Humans construct robots to do tasks that people do. AI works by collecting data and processing the data with an algorithm. The program looks for patterns that repeat and uses those patterns to predict the future. It bridges the difference between machines and humans. It's a computer or machine programmed to do what humans do. It's teaching machines to do human activities faster with fewer mistakes.

Artificial intelligence includes:

- machine learning
- facial recognition and identification
- chatbots
- epayments
- natural language processing

It starts with a goal a human sets, collects data, and applies a procedure. The method or algorithm changes to accommodate the new data as the data changes. Eventually, the process reveals a pattern that leads to a result. The program is constantly checking if it's correct. In the case of facial recognition, it may check 1.4 billion times in a second to determine which face corresponds with which identity. If it fails, it must go back, try, and forget until it finds a solution or remains stuck until enough data reveals another process to explore.

What can AI do?

AI is applied in many ways in government and business settings. Beginning in the early 1960s, manufacturing used it to automate activities that were too dangerous for people to do. Since then, it has been applied in many ways to improve what machines can do. In China, facial recognition technology allows people to withdraw cash, check in at an airport, and pay for goods. AI can read people's emotions. In the US, computers can analyze data from an X-ray and determine what problem occurs in someone's lung. The technology can do it as well as a trained medical professional. Robots build cars faster and better than humans and can drive a vehicle with fewer mistakes than people. Spell checkers can find errors quicker and better than the best proofreader. Outside of being human, I doubt there are many things humans can do better than AI. Yes, it takes time, energy, and money, but the result is typically way beyond what humans can do.

Facial Recognition Technology (FRT)

Identifying people has enormous potential for efficient control of the population. If you can identify one person out of 1.4 billion in a second or 2, you can gain many advantages. You can collect data for each person and input that data into data for 1.4 billion people. FRT gets rid of paperwork and ID cards. No one needs to check your identification. You don't need to remember to bring your ID card. Debit and credit cards and endless hassles are avoided. No one can steal your identification. This eliminates mountains of paperwork and makes your economy more efficient. People save a lot of time, energy, and money. With data from your profile, the government can tailor solutions to your particular needs. In China, FRT is used extensively and has raised alarms as it could be used in ways that violate human rights.

Data Collection

Data is the raw material of AI. Some are relevant, and some are not. Data is collected to recognize a pattern. Sensors collect data, which an algorithm processes and turns into a solution. Data are the gold of the AI economy. It can be used to predict what people will do. Algorithms assemble and process data. The program recognizes repeating patterns and builds models to predict people's actions. Computers can use data to learn how to solve problems.

Data Mining

Data is money, or so some people say. Am I searching for something on the internet, or is the internet searching me? Both things may be true. Can you control your spending? Would you spend money if you weren't using the internet? We are bombarded with commercials day and night. Capitalism demands it. Data mining is extracting information about what you like, what you buy, and what you want more than anything else. Data mining collects information about you when you search. The words you type into the search box of the search engine are called keywords. Often, these are the beginning of a search for something to buy.

I need a winter jacket, so I type "winter jacket" into the search engine. The first thing I see are ads on the top and the side. If you're unfamiliar with the process, click on the top one with just the right words to get you interested. All seems perfect, but if you are on a budget, it becomes apparent that results on top of the search results are too expensive perhaps double or triple the regular price. For some people, it doesn't matter. They have lots of money and will pay whatever. The function of data mining is to accumulate enough data on someone to predict what they will buy next. Once this has been done, making a sale could be easy. Data mining can pay off well in the long run, even if people don't buy much. If a hundred thousand people are buying not much, that can add up to a whole lot. Once Google has the infrastructure set up, running the search engine per person costs a small amount. Also, making free programs pays off because it's all about traffic and trust. If I give you something valuable for free, will you like me? How about if I do it over and over and don't charge a penny? The internet is like having all the libraries in one spot you can access without leaving your chair.

Sensors

Sensors are used to collect data, which are processed by an algorithm that changes based on the new data. AI is used to automate all activities of the economy. It started in 1961 in manufacturing and soon after food processing. AI is spreading rapidly to automate all human activities to make money. Sensors collect data. Sensors can detect changes in pressure, temperature, motion, elevation, position, optic, flow, gas, chemical, torque, force, gauge, current, voltage, and velocity. The machine collects the data, processes the data, applies an algorithm, and produces a result. Businesses are transformed to use AI to reduce the time, energy, and money required to make something or provide a service. The sensors collect data, fed into an algorithm that recognizes patterns that can lead to predictions. Based on recurring data patterns, AI determines which event has the highest probability of occurring next. Sensors accumulate data, which is a critical component of AI.

Machine learning

Machines have just recently been able to learn. Some are programmed to do one function and then again to do another. This is not machine learning. The machine does not change when the conditions change. If you want the machine to choose a red apple and it chooses green apples, it's not a machine that learns. A machine that learns can compare the color of the apple to the desired color and choose the desired color. Machine learning develops algorithms that change as the data changes. Data accumulation and selection is the first step in the process. Quality of data is a critical component of the process. With relevant data, the results will have value. The algorithm processes the data once selected, processed, and culled for incorrect data. After data is selected, a person or machine may make rules for processing the data. Sometimes, they use a person first and a machine later. The machine learns by comparing new data to stored data to determine if it's significant. With testing the data, it will be known if it is good. Machine learning is an AI process that accumulates data and applies an algorithm, and as the data changes, the algorithm changes. Machine learning makes predictions from patterns it recognizes in large sets of data.

A logic of machine learning:

- a job is several tasks
- the task determines relevant data
- sensors measure data
- an algorithm processes data to make the output
- as the data changes, the algorithm changes
- machines learn when an algorithm changes as data changes

Distinctions between humans and robots

People invent machines and computers to advance their health. Steel rods hold backs straight, screws hold bones and muscles in place, and pacemakers keep hearts pumping correctly. Another threshold is about to be crossed: the link between mind and machine. Neuralink is an invention by Elon Musk to embed a chip in the top of the skull. The surgery is minimally invasive. You go in the morning and leave in the afternoon. A robot does it. The device can be removed without damaging your brain. It's recharged at night with no wires. More than one device can be implanted.

What do they say Neuralink will be able to do?

- restore body motion to people with spine injuries
- restore vision to blind people
- reduce pain
- prevent illness
- solve mental illness

In the future, Neuralink may store, restore, and install memory in a robot or human body. To insert Neuralink, a robot inserts 1024 electrodes into the skull. No general anesthesia is necessary. The procedure is done in less than 6 hours. If large-scale use becomes popular, the company hopes to make it available for a few thousand dollars. They are confident that it can download and upload human memories in the future. It may be possible to download your memories and abilities to another person or robot.

The limits of AI, robots, and drones may be impossible to predict. In 10-20 years, imagining how we could live without this technology may be challenging. It may increase inequality as wealthier countries have access to it, and impoverished countries languish in nonstop global warming disasters. Most applications of AI may have people in charge. Robots and automation perform better than humans at nearly every task.

Machine learning is a quantum change maker.

Until recently, machine learning has never been possible. A machine can do functions without humans controlling it. With built-in sensors to collect data, the sensors replace human eyes, tongues, noses, and ears. Sensors can detect data that humans can't. Since the data is measured, it's more accurate and valuable than the data humans collect. We have rough ideas about the wind speed or the temperature, but it's more a guess than what a thermometer tells us. Human sense information is less accurate than a sensor. Also, humans can't store, retrieve, and process data like machines. Once set up, the advantages of a machine over a human are many and significant. The machine collects, organizes, and processes data to meet a goal set by a human. This change generates many consequences for a large economy based on consumption. Machines can do most of the production work that people typically do. In developed countries, it's rare for people to wash clothes by hand. As it is applied to the economy, many people will be pushed out of their old jobs and into the AI economy.

Machine learning is a leap that hasn't been done before and takes machines to another level of power and practical application. Before this, computers were designed to perform operations; now, the algorithm perfects the operation, makes mistakes faster than humans, and advances beyond those mistakes faster. With enough data, machines can learn to do tasks without human instructions or control.

Governments can use AI to meet needs sustainably.

The function of government is to provide a decent living for its people. What does that mean? Governments can prioritize providing healthcare, housing, and food in ways that enable the long-term viability of the environment, air, land, water, plants, and animals. Businesses are very inventive. When you set limits for businesses, which is what they want and need, they can function without being forced to close because climate-change disasters make business impossible. Unnecessary functions support the necessary functions. Republicans love this. As meeting needs are satisfied, companies can produce more movies, fancy paint, and pretty dresses. As the supply of necessary items falls, they can create more of what is required. What could make a capitalist happier? It's supply and demand.

Implementing sustainable transport with AI will advance a country.

As AI expands, machines will do more of the work done by humans. People will likely exercise less. Can infrastructure help people live well? Sustainable infrastructure expands the chances for someone to exercise. One goal is to make walking, bicycling, and using transit the most pleasant, efficient, and inexpensive way to get to and from destinations.

AI machines can be used to run buildings and cars without the use of fossil fuels.

Machines controlled by artificial intelligence enable advantages to make the country run without accelerating climate change. Electric and hydrogen power may be the power of the future as fossil fuels become extinct instead of people. Heat pumps and fuel cells may replace fossil fuel boilers. `

AI can curb the cancer of consumption.

Far and wide, the Western world races to accelerate climate change. The capitalist will to profit is the fast track to extinction. Like a child without limits, the corporate charade is racing the world to climate change extinction. AI could be used to minimize the production of that which destroys our means of life. Why should people use AI to march to suicide? Do we all need so many electronic toys or 20,000 different pants and shirts? Ever look at what people throw away? The wealthy throw out enough to run another country. Why is that so important? The endless stream of waste is a measure of the economy's inefficiency. Sensors on garbage trucks could measure what people throw out that could be recycled.

Robots

Robots are computer-controlled machines. Robots with many components and complex functions require huge outlays of time, energy, and money to develop, produce, and maintain. Robots may operate with or without human control. Most robots have a body made of metal or plastic, motors, and a computer. Constructing a robot may take years with a large team of engineers and scientists to design the right size, shape, and material for the components. Robots that work in extreme temperature and wind conditions or require special functions may use rare earth elements, which may be very expensive, challenging to mine and process, and unavailable. Robot use is expanding exponentially.

Robots are used for the following work;

- any activity that can be codified
- in medicine, do surgery, disinfect a room, provide therapy, and transport items.
- driving vehicles
- vacuuming floors
- information gathering, analyzing, researching

- packing, showroom, and warehouse moving

Robots for medicine

Robots are used to advance medical care. The DaVinci system empowers doctors to operate in tight locations in the body using a robot with 3d vision and minute-cutting devices. This technology provides more control and is less invasive, so patients recover faster. The Tug is a robot that moves materials and supplies around the hospital. With sensors that detect people, it can ask people to step aside as it makes deliveries. Xenex is used to reduce hospital-acquired infections and reduce labor. It cleans a room quickly using pulsing ultra-violet rays that destroy a room full of germs in minutes. Paro is a therapeutic robot that does some of the functions of a pet and is suitable for patients with dementia. Paro responds to human interaction and tailors itself to the patient. It reduces stress and anxiety as it winks, flutters, and makes sounds. The CyberKnife operates in sensitive areas like the neck and prostate. It removes tumors that doctors can't take out with surgery. It doesn't damage surrounding tissue and is non-invasive, allowing patients to get surgery with minimal side effects. The Neurolink robot is another quantum leap in technology. It may enable paralyzed patients to use their bodies usually, reduce pain and mental illness, and, eventually, allow patients to download their memories to a robot or another human.

AI for farming

Farming may seem like an art, but AI can do it more efficiently. Farmers can use AI to:

- analyze the soil to determine how to fertilize the soil to grow healthy plants
- minimize the amount of fertilizers, herbicides, and pesticides to use
- identify plant diseases, pests, and plant nutrition to use on the farm
- help farmers make decisions by analyzing weather patterns over time

- use drones to collect and analyze data over large areas of crops
- use AI to plant and harvest crops more efficiently.

Like most other areas of human activity, AI can get the job done faster with less waste of time, energy, and money.

Drones (Unmanned Arial Vehicle)

Drones are flying robots with various functions for farmers, doctors, governments, and others. Farmers increase the production of plants and animals. Emergency responders and doctors use drones to respond faster, delivering better health care. Governments use drones to gather information to act quickly to help people or minimize the consequences of fires, disasters, and emergencies.

Agriculture use of drones includes:

- survey land for problems with soil, crops, and livestock
- deliver precise amounts of pesticides, fertilizer, and herbicides
- pollinate tree crops to increase yields up to 40% or more
- spot diseases or nutrition problems with animals
- plant trees six times faster than humans
- plant seeds and crops

Medical use of drones includes the following:

deliver drugs, supplies, and equipment to emergency areas
Search for and find people and animals in rescue operations
to survey disasters and areas with biological and chemical hazards
send live footage of areas where people need emergency services

Governments use drones to:

- track the movement of people at demonstrations and large gatherings
- coordinate attacks on civilians who are exercising their human rights
- survey to plan, coordinate, and deliver disaster relief
- gather data to assess problems and deliver solutions

- collect evidence at a crime scene
- survey and manage traffic
- coordinate lifeguard operations

Business use of drones

- deliver packages, equipment, and supplies
- deliver food and drinks in bars and restaurants
- use cameras to survey operations
- automate human activity efficiently with less time, money, and hassle

In real estate, drones are used to:

- make photos and videos of land and buildings from the side and above
- capture visual data to make an interactive aerial image similar to a satellite view
- in-depth, high-resolution photos of terrain, land, and building
- manage and maintain properties

Drones are used in the construction industry to:

- inspect the land, outside and inside of the building
- used for security to reduce theft
- used to increase safety

Weather Forecasting

Drones are used in weather forecasting to make forecasts more accurate. Weather drones use temperature, humidity, and wind sensors to gather data at the lowest level of the atmosphere. Algorithms process data. The meteorologist interprets the results to determine the forecast. Drones are useful in coastal and mountainous areas. Drones collect information in regions that are too dangerous for people.

Drone laws

Federal, state, and local authorities regulate drone use. Governments don't want drones flying drones too close to people, buildings, water bodies, and wildlife areas. Drones must not interfere with airplanes, helicopters, or emergency responders and people. Don't bother anyone else's privacy, buildings, or plants and animals; you are probably going in the right direction. You must be able to see your drone when you are flying it. One set of laws applies to recreational use, which is narrowly defined, and commercial use, which is more restricted and carefully regulated. Drones without enough battery power or fuel will fall. If you're not cautious, you may hurt a person, plant, animal, or building, and you may be responsible for the damages.

Robots for Construction

Robot use for construction is expanding. In the US, workers would instead do other things because of the many challenges of construction work. To use a robot on site requires careful preparation. Some robots can build all the exterior and interior walls in a couple of days. The building department must approve work before starting. Some cities do not allow robot construction, some have approved robot-constructed buildings, and companies are rapidly expanding their use. Robots can build high-performance, net-zero buildings that require little maintenance and provide more protection from extreme weather than conventional construction. Robot houses may cost 20-40% less than traditional construction. Now, it's common for the walls to be built by machines and the rest done with non-robot methods. With higher performance and lower costs, robot-constructed houses may expand rapidly.

Autonomous vehicle robots for heavy digging, excavation, and pushing

Autonomous vehicles excavate earth for the foundation of buildings, dig trenches, level soils, load trucks, and push heavy objects. They may do their work at night, which can accelerate the pace of construction. Autonomous vehicles may operate during periods of extreme heat and cold.

3D House printers

3d house printers lay thick, dense layers of concrete, wax, and foam polymers. 3d printers may be used for commercial and residential applications. Non-straight, poly-angled designs are more accessible and faster with a 3d printer. The walls may be more airtight, which reduces heat loss on windy days. Most printed homes may be built to higher energy standards than other homes.

3D printers construct sustainable housing

3D printers can build low-cost, sustainable housing, especially in areas without freezing temperatures. These robots build high-performance buildings that protect from rain, wind, and hurricanes. The house can be constructed for up to 40% less time and money. One development in Tobasco, Mexico, used a 3D printer to build 500-square-foot houses with a Vulcan II 3D printer. After 24 hours, the house walls are printed. Carpenters add the doors, windows, and roof to finish the project. Financing is included without profit or interest. With payments of $20 per month, the housing is long-term and affordable. Plans are to build 500 units of this housing.

One task construction robots

Construction robots that do one primary task offer advantages and have significant limitations. One-task robots may be easier to build, cost less money, and be easier to integrate into a construction site. They may use a track with an arm that uses precise movements to build a brick wall, finish gypsum board, or weld metal parts. One task robots may do more work in less time do it more consistently with fewer mistakes, but may be too expensive for smaller operations or small jobs,

Single-function construction robot use is becoming more common to do limited tasks, including the following;

- hang and finish gypsum board with an HRP-5P, which is a robot manufactured in Japan
- tie rebar on roads and bridges
- lay bricks using HadrianX
- weld machine parts

Robots are used for work that humans are not so good at, including;

- heavy, monotonous work that requires precise positioning
- work at night during extreme heat or cold
- exact work that humans can't do

Low-cost sustainable roofing made with discarded plastic

Resilient Modular Systems is a company that designs roofing for low-income people worldwide. The idea is to use discarded plastic to print roofing panels that are easy to install and maintain. The panels are 4' x 4', interlock, and are easy to install and maintain. The panels are set up to replace corrugated steel roofing that rusts quickly, requires frequent replacement, and generates too much pollution.

AI for making buildings

Some predict that by 2050, robots and machines will do most of the construction work. Buildings will be designed for robots to build.

AI for transport

Shenzhen buses are self-driving and have been operating since 2017. Shanghai has roads set up for self-driving vehicles.

AI is used to help bike riders get to work, school, or do errands efficiently.

The Bico system is a software program that organizes city bike-sharing operations to make them more efficient. City bike-sharing programs operate in cities around the US, including New York City, Chicago, and Washington DC. Users pay a yearly fee and ride as the program specifies. In New York City, for $100-200 a year, users can use the bikes without doing any maintenance. One problem with the system is that riders from one area will use the bikes to leave the area and not return till after work. After the morning rush, there are no more bicycles to use. The software organizes the system to reduce these shortages. It coordinates drivers to take the routes with the most minor hazards, traffic, and safety problems. The software is a tool to optimize the use of bike-sharing systems to make cities more sustainable. Bike riders can access more bicycles and have a safer, more efficient bike ride. AI functions to help bike riders who depend on bikes to get to work, school, and run errands.

Transit users can ride autonomous vehicles to transit in Arlington, Texas.

Transport uses of AI depend on data. With data from transit, bikes, and pedestrians, machine learning can optimize sustainable transport infrastructure and make transit more efficient. One system in Arlington, Texas, integrates on-demand autonomous driving vehicles to give a quick ride to transit. The vehicles operate within a small area. It's a pilot project to determine if this can be expanded to other cities. Students ride free, and a computer takes data from the bus rapid transit system and autonomous vehicle to maximize the system's efficiency. The limited scope of the project makes it more feasible. If a city-wide bicycle system with sensors is included, the system may offer a compelling alternative to auto use. A question remains. Why not walk or ride a bike to transit instead of using a vehicle? This may increase the number of people riding transit. Only some people can walk or ride a bicycle. However, it may encourage those who can walk and bicycle to use a vehicle, which is worse than riding a bike. More auto use, even electric, is worse than walking or riding a bike to transit. Data is collected from bicycles, autonomous cars, buses, and traffic lights and put into a computer, which uses algorithms to maximize the efficiency of the transport system.

Scientists use AI to process data,

In the study of genes, AI is used to process and analyze data to predict the structure of proteins. This information can help identify and treat diseases. Environmental scientists are using AI to predict weather changes to help cities adapt to climate change. Astronomers in the Kepler mission used AI to process data to find Earth-size planets. Conservation scientists interpret data with AI to monitor populations of seals that would be very difficult to count by hand. Organic chemistry scientists use machine learning to model and predict processes in plants. Particle physicists use AI to find patterns in neural networks. For scientists, AI is a tool to advance their ability to test their ideas and find solutions.

Trikala, Greece, implements automation.

A city of 82,000, Trikala, Greece, is a city of innovation, design, and artificial intelligence. In 2014, the town was in a dismal economic depression. With the help of a new mayor, the city worked to apply automation and machine learning to improve the efficiency of transport, healthcare, and city services. By installing sensors for parking, traffic signals, water use, lighting, and other functions, the city was able to accumulate data for algorithms to deliver public services in less time while improving the lives of the poor and elderly with a system of health care. Maps, charts, and ATM-like equipment at city hall allow people to print receipts that they've paid all their bills before transferring property.

The city installed sensors in garbage cans, lights, water, electricity, and 5g technology. Transport is controlled with sensors in traffic lights. The city set up a pilot program to help the poor and elderly get care more efficiently by using a health monitor. AI is used to:

- finds parking
- controls double parking
- control parking in handicapped areas
- collect parking fees

Federal laws regulate drones in the following ways

All drones for recreational use must be registered on the outside of the drone.

Drone users must carry proof of registration when flying.

Drones weighing greater than .55 pounds and less than 55 pounds must be registered.

Flying over 400 feet is forbidden.

States and local municipalities may make laws with more restrictions.

Some states may prohibit:

- use of drones to record audio or video information of other people
- locating, harassing, or hunting animals with a drone
- flying too close to a government facility, including schools, hospitals, and water facilitates
- using drones to fly close too oil and gas facilities and operations
- cities and towns from making more restrictive laws.

One function of government is to sustain the economy.

AI is quickly transforming the way society functions. AI machines will take over many jobs that involve manual labor, routine clerical work, and most of the economy. Most human work activities can be done faster, cheaper, and more consistently by AI machines. AI is moving full steam ahead, sometimes with little government coordination, regulation, and direction. While there's a slight chance that all will work well, the facts lead to another conclusion. With AI machines automating the economy, many will be out of work. Others will need to be trained for the AI economy, which is more about making the conditions for AI to accelerate to produce whatever people will buy than a coherent plan of action to meet needs sustainably. Another approach prioritizes the use of technology in a way that addresses the critical problems here and now. Meeting needs sustainably in the community can use our resources efficiently to avoid endless global warming catastrophes. Prioritizing sustainable transport and development over a blind rush to produce anything that can make money will put the country on a positive path that will avoid endless global warming disasters.

CHAPTER 21
JOBS AND THE ECONOMY IN THE ERA OF AI

The economy is in motion—global warming pressures governments to power buildings and vehicles without using fossil fuels. AI expands capitalism. A corporation that successfully automates will have an advantage over the competition and can lay off workers. Using large sums of money, corporations develop AI-controlled machines to get work done faster with more consistency and less money. Robots don't have human rights or need insurance or a pension. Designing, building, and maintaining robots is a tremendous amount of work requiring large sums of money. It's not a project for a person or small group. Work must be set up correctly for a robot to function. Some of this work must be done by humans, some by machines. Half of the jobs will be replaced by automation. The economy pushes people to adapt as the transition to maximize automation pushes through the economy. A lot depends on the genius of designers at the top, who started uniquely and developed their ideas most completely—capital pressures innovation to produce more efficiently.

Jobs are dumbed down so they can pay less and eventually automate with AI-controlled machines.

Running a business is about making money. Employers work to

minimize labor costs while increasing the price of what they are selling. If an employer can reduce labor costs, they can make more money. Higher-skilled workers are paid more because they can work without supervision, do many different tasks, and produce higher-quality work. If you made a machine to do the skilled work and used unskilled labor to set up the machine, you would have a larger pool of workers and more potential to keep wages lower. Highly skilled workers are hard to replace and can negotiate higher salaries. Eventually, the skill level will be low enough for machines to do more work until many or most jobs have been eliminated.

Companies justify using AI when there's a "shortage" of workers.

When Donald Trump was elected, he stopped letting people into the country from south of the border. Many of these people have had to live in desperate economic situations caused by war, economic dislocation, and environmental destruction. After living a difficult life, many are motivated to work and live in the US, where the economy produces higher-paying jobs, and the standard of living is higher. These workers do jobs that most people from the US like to avoid. One justification for AI is a labor shortage. We need AI because there aren't enough workers to get the job done. For years, pay for construction workers has been eroding. Wages don't increase with inflation. Construction workers faced job hazards and physically demanding work as wages decreased as immigrants were willing to work for less. In 2021, the demand for single-family homes has dramatically increased. Shortages of common materials drive up prices. Worried about inflation, investors move to increase their holdings of real estate. AI construction robots can do the work faster and cheaper, but setting up these machines requires a lot of investment. Changes in the supply of workers are used to justify using AI.

How the government and the internet took over expediting jobs

I lost my job in 2008 when the market crashed, and construction work vanished. With a bit of training, I was doing expediting work. Expeditors used to take plans and paperwork from architects and bring them to the Department of Buildings in New York City (NYC) to get approved. Expeditors had to wait a long time in line, sometimes hours. If the architect made one little mistake, it would have to be taken back and done repeatedly. The clerks behind the windows probably lost their jobs, too. It was an old, backward system. The government automated those jobs out of existence. There used to be a huge room filled with expeditors at the Department of Buildings in New York City. All these workers were hustling to make money doing this job. Within ten years, most people working that job were automated out of a job. Most of the work could be done electronically with fewer mistakes, requiring much less energy. Most of this work can be done electronically, so why not? It's a massive waste of time and money. Yes, it's a job, but it's a dumb job. Most of the time, I was sitting there waiting. Many jobs are like this. So, I found another job that was supposed to pay a decent income but riddled with a roller coaster of hurdles.

AI, the internet, tenants, and landlords have replaced real estate rental agent jobs.

As the expeditor work faded, I looked for another way out of the economic malaise. I got my real estate license, found a broker to work under, and began work as a rental agent. My job was to help people find an apartment in New York City (NYC). Busy people who don't know what's available need help finding an apartment. Some make enough money to pay a fee equal to one month of rent. Other tenants who are not familiar with the system are shocked. Before the internet, looking online and finding ads for thousands of apartments with a click of a button was impossible.

There were ads in a newspaper and apartment listings that only real estate agents had access to. In NYC, large, busy landlords use real estate agents to find tenants to rent their apartments. Because the work can be territorial, starting can be difficult. Also, you only get paid if the landlord accepts the tenant and the tenant pays the one-month rental fee. I didn't have a firm connection with a broker, so I worked for many months and have yet to get paid. There were many other real estate agents doing the same thing. It was very competitive, and because rent is so high, most tenants want to avoid paying an agent one month to find an apartment. At that time, landlords had been listing their apartments, and some found tenants who needed a real estate agent. Eventually, the New York State government passed a law that said you can't charge a tenant a fee to find an apartment. The landlord has to pay the cost now, which many tenants probably like. The government wanted to stop real estate agents from charging these hefty fees to find an apartment. The internet made it easy to list an apartment and easy for a tenant to locate an apartment. The costs were very high and only made sense if you had a lot of money and didn't mind paying to support a real estate agent's income. Most tenants may prefer this because it makes apartments in NYC more affordable. Landlords who have the better side of this may not like the new system. Many real estate rental agents have lost their jobs and had to find other work.

Clerical worker jobs are on the AI chopping block.

As computer use expands and companies find more accessible and faster ways to do things, it becomes evident that a job is unnecessary or can be done easier and faster using a computer, machine, robot, or drone. Paying money with a check, then mailing it, then having a clerical worker open the letter, inspect the check, record the payment, and submit it to a bank is a system complete with too many chances for errors. All this can be done safer, faster, and more accessible than people walking to the post office with cash, purchasing a money order, mailing it, and hoping it arrives on time. If the person sending the letter has terrible handwriting, forgets to write the address correctly, or makes a small mistake on the money order, they can lose a lot of time and money. Also, if they carry cash around, they could get robbed, which is very dangerous. If a thief sees you going to the post office every month, buying a money order, and stuffing an envelope, they may target you, and the results could be terrifying. So if the money order does arrive, the clerk at the government office, bank, or business has to open the letter, put away the money order, and take it to the bank, hoping the bank will accept it and deposit it. If paid electronically, there would be much less work for all involved, and it would be safer.

Can a machine outperform a doctor or plumber? Really?

Some people think an artist, a doctor, a plumber, or someone else can do a job better than a machine. People have lost that battle every single time. Can you dig a hole faster than someone with a shovel? Of course not. Can you dig a hole with a shovel faster than someone using an excavator? Of course not. It takes a lot of work and energy to make and set up the excavator, but now it's expected. Companies know how to make them, and boom, just hand over the money, and you've got an excavator. It can do the job of many men in much less time. It's more accurate with a computer controlling it with a Global Positioning System (GPS). So, what jobs can be automated? With enough time and technology, any job can be automated, and once they are automated, the machine will be doing work that humans can't do, and they do it faster with more consistency. Ever try to beat a calculator?

Just push the art button.

Some people say an artist's job can't be automated. Bill Skodje said, "When I need to get work done real fast, I just click the art button in the Photoshop application." He was joking, but the point is made. What is going on is that corporations are choosing what activities to automate. Now, the distinction is how much will be automated. They start with what seems possible and increase until it's mostly machine work. Some people say doctors will be put out of work. Yes, some doctors will, but they do many different things, and we have yet to arrive. Eventually, machines will do doctors' work better than a doctor. When it's obvious a machine can do the job better than a human, it's on the chopping block. I've had two jobs that were automated out of existence within ten years. After 2008, there were no construction jobs.

Work for humans will shrink.

Some reports estimate that 50% of the workforce will lose their jobs to automation. Work for people will decline as AI, automation, and machine learning expand.

Jobs will decrease in the following areas of the economy.

- banking
- retail and e-commerce
- manufacturing
- marketing and sales
- supply chain management
- logistics and manufacturing

- gaming and entertainment
- construction and development

Jobs that may avoid automation

People who work in very unique areas of the economy may have secure jobs. If only a few people use what you make, it may not make sense for a corporation to invest a lot of money into something that will sell a few units. But if your job costs the employer too much money and there is an automated way to do it, the employer may go that route.

What jobs will increase when AI expands?

With quantum leaps in technology, new jobs are created. With automation sweeping most sectors of the economy, people will be needed to design and construct computers, robots, drones, and many types of automated equipment and machines. Humans will supervise many computers, robots, and drones as they operate. Some, like AI-controlled vehicles, will not require a human to operate. As automation reaches another level, robots may design and construct robots. Automation and machine learning will require scientists, engineers, designers, and directors to automate the economy.

The following is a partial list of jobs that will increase as AI expands:

- Machine learning researchers
- AI engineer
- Data mining and analysis
- Design Engineer
- Robotics engineer
- Data scientist
- Business development specialist
- Product technician
- Sales manager
- Algorithm engineer
- Machine learning engineer
- User experience designer
- Software engineer

The demand for people to make robots increases the number of jobs for teachers and engineers.

As technology changes, so do the needs of employers. More engineers, designers, and innovators are needed to automate economic functions. Making robots is something most people have never done before. Some companies teach people on the job.

AI Economy

Many jobs are generated by the push to automate human activity. Who will make the robots? Can robots do it? Computers, machines, and sometimes robots are used to build robots, but without humans, there would be no robots. People are needed to create, operate, and dispose of robots. Who will buy and set up the robots and ensure they do their jobs properly? Inspection robots, perhaps the ones that can tell what you are thinking, even before you. The intelligence of machines has already eclipsed what humans can do. A giant leap is when machines can build machines to advance the interests of machines. When machines do most of the labor, what will people do?

As AI continues to transform the economy, certain job roles are expected to expand due to the increased demand for human skills that complement AI technologies. Some of the job roles that are likely to expand include:

1. Data scientists and analysts: Professionals who can interpret and analyze data to derive insights and make informed decisions will be in high demand as AI generates vast amounts of data.

2. AI trainers and explainers: Experts who can train AI systems, ensure their ethical use, and explain their decisions to stakeholders will play a crucial role in the AI-driven economy.

3. Cybersecurity specialists: With the growing reliance on AI technologies, the need for cybersecurity professionals to protect data and systems from cyber threats will increase.

4. Human resources and talent management: Professionals who can manage the human side of AI implementation, including workforce planning, training, and employee well-being, will be essential in the AI-driven economy.

5. Creative professionals: Jobs that require creativity, innovation, and emotional intelligence, such as artists, designers, and content

creators, are expected to remain in demand as AI cannot replicate human creativity.

Overall, job roles that involve complex problem-solving, critical thinking, creativity, and interpersonal skills are likely to expand as AI takes over routine and repetitive tasks in the economy.

Jobs and the Economy

Many jobs will be lost; it's a correction from the past. People use AI to develop and produce machines, computers, robots, and drones to automate people's actions. The economy is going through another phase of technical development. Some say it's similar to when electricity became common. Most of it will be done to feed the pockets of the super-rich as millions are left without work, housing, healthcare, and education. With training and perseverance, many can study, train, and make AI that advances a movement to sustain transport and development.

CHAPTER 22
LIMITS OF AI

Limits of AI

It's easy to imagine a robot or machine working faster and more consistently than a person. Examples include manufacturing a car, processing food, and washing clothes. Spell checkers do better than the most intelligent humans. One fear is that robots will make robots, which could eliminate humans, or a more likely possibility is that robots will do most of the work humans do, and there won't be any work for humans. In most cases, people control the robots. A person wants the floor cleaned; they turn on the robot, and the robot cleans the floor. It's easy to stop the robot once you know how. Robots can only get the power they need to keep going without humans if they are solar-powered, but what if they get stuck in a building without light? At this point, robots perform limited functions, but whenever people think of something a robot can't do, a person finds a way to solve that problem.

10 Limitations of AI

1.AI must have large amounts of data. How the data is collected and processed may reflect the biases of the person, institution, or society collecting it. Do the sensors measure what's important? Can the sensors measure the right amount?

2. Bias may be inherent in how the data is collected, coded, and selected. An algorithm may not identify and remove the wrong data without a way to cull the unfair or biased data. The data could lack social context and could spread prejudice. When algorithms are tested only for speed, the dominant bias, typically bigoted, may taint the results in disturbing ways.

3. Processing large amounts of data can take time. Companies with more money and resources may do it without concern that it's too expensive. However, the issue for smaller companies may prohibit the operations.

4. There needs to be an overall agreement about what is OK and what is not. There's Elon Musk, putting chips in pigs' brains and displaying them for humans to look at them. We don't know what happened to the other pigs or animals with chips inserted in their brains. What will happen if they figure out how to download a human brain and put it into another person?

5. Who is liable? What if markers are set up on a highway to guide a vehicle, and the marker is moved to the other side of the road, and the automated car causes an accident?

6. Collecting and processing data can use large amounts of resources. Bitcoin is one example. The longer it's available, the more energy it consumes. It's almost like quicksand. People are making this stuff, and it gets bigger and bigger. Sometimes, when this happens, it's hard to stop it. Have you ever tried to clean up your room and throw stuff out? In the US, it's easy to accumulate too much. Whatever novelty suits your fancy becomes an addiction to get that momentary thrill of possession. An infrastructure of people, ideas, events, and dependencies is generated that people may want to continue. It gets wound up like a giant flywheel that takes years to start and years to stop.

7. AI requires supervision. Will semi-autonomous vehicles make it easier for people to fall asleep from boredom and cause more accidents? If someone has to watch out for AI mistakes, what's the significant advantage?

8. AI machines often do one function. Most are not able to do several tasks. If you tried to make a machine that does all human functions, it would take a lot of space and not be practical.

9. AI generates pollution. Imagine piles of broken, poorly designed, or worn-out robots. Where will they end up? If they are intended to be reused, renewed, and recycled, that could reduce the problem. In New York state, some municipalities charge $20 to throw out a TV. Ewaste contains lead, mercury, beryllium, cadmium, and brominated flame retardants. Clever companies harvest and use it. When the equipment is broken, these elements may seep into the environment.

10. AI may be difficult to maintain. In the process of making AI, errors may accrue that lead to bias. Data must be selected based on the goal you want to achieve. If the goal is to make money, the goal is not to be fair. Making predatory loans may make more money, but it also makes victims, and that's not fair.

11. Algorithms can be biased based on the algorithm makers' social, political, and legal biases.

AI mistakes, some with consequences

In the push to profit and under pressure to produce, designers and engineers who don't live with the consequences of their mistakes.

Facial Recognition Technology (FRT) can generate big problems. One example includes the use of technology to identify criminals. FRT wrongly identified star athletes as criminals. The players were not white, and the mistake generated media from far and wide.

In 2016, A Microsoft chatbot named Tay was in a conversation and mistakenly spouted out Nazi-like epithets. Although it was taken down shortly after, this example shows how making this technology work properly is more challenging.

Algorithm Bias

Systematic and repeated errors in algorithms can cause bias on a large scale. The problem could be in the design of the algorithm, unintended use of the algorithm, or decisions made about how the algorithm codes, collects, and processes data. Search engine results and social media platforms violate privacy and reinforce biases of race, gender, class, and sexual identity. Widely considered neutral and unbiased, algorithms project more authority than people, and who do you blame when an algorithm puts out the wrong information?

The bias of the algorithm

When data is collected, assembled, and processed, it may generate errors because of the way humans choose the criteria to catalog the data. As programmers assign priority to the data or assess and sort the data, the programmer's bias may be present in the output of the data. Algorithms set up to collect data that use criteria chosen by a person may produce the bias of the person who decided the requirements. If the algorithm uses data people select, it may continue to be biased toward those selected to submit their data.

Algorithms project a machine-like authority and may be used to avoid responsibility for the outcomes. When people project authority onto the output of a website or AI technology, they may not consider other options, interests, and desires. Even when an algorithm is well designed, it can still generate problems, but these may be difficult to trace and fix as algorithms go through various production processes. Decisions made by a group of designers may be obscured in the code and forgotten as the years fly by or the code changes.

Many problems can arise if the algorithm isn't adequately tested in various applications, including unintended uses.

These answers are from an artificial intelligence platform copi ai and chat Gpt. What are some limits of AI?

- Bias and discrimination within AI systems1
- Lack of transparency and explainability in AI decision-making
- Difficulty in understanding context, meaning, and nuance1
- Limitations in creativity and imagination compared to humans2
- Dependence on data, which can be limited or biased
- Vulnerability to manipulation or hacking
- Inability to handle unexpected or novel situations
- Ethical considerations around the use of AI1
- Cost and complexity of development and implementation2
- Potential risks to job security and the economy as a whole1

AI Limitation: Bias and discrimination within AI systems can arise from many sources, including biased data used to train the AI system, the unintended biases of the programmers designing the system, or the way in which the AI system is deployed or used. For example, if an AI system is trained using data that underrepresents minority groups, it may produce biased results when those groups are later used in decision-making processes2. Additionally, even when AI systems are created with good intentions, they can still produce discriminatory results if they are not designed to account for the full complexity of the social and ethical considerations involved in making decisions and implementing policies1.

Furthermore, the limitations of AI systems in areas such as context, nuance, and creativity can compound the effects of bias and discrimination in ways that deepen existing inequalities, rather than reduce them1. For these reasons, it is important to ensure that AI systems are carefully designed, tested, and monitored to minimize the risk of bias and discrimination, and actively work to promote equity and inclusion in the decision-making process.

AI Limitation: Lack of transparency and explainability in AI decision-making

The lack of transparency and explainability in AI decision-making refers to the challenge of understanding why an AI system arrives at a particular decision or recommendation. In some cases, AI algorithms can be complex and difficult to interpret, making it challenging for humans to understand the underlying factors and reasoning behind the system's output1.

There are various factors that contribute to this limitation:

1 Black box problem: Some AI algorithms, such as deep learning neural networks, operate as black boxes, meaning that even the developers of the AI system may not fully understand how the decision-making process occurs. The complex interactions of thousands or millions of parameters make it difficult to discern the specific reason behind a particular decision2.

2 Lack of interpretability: Many AI models generate predictions or decisions based on patterns and correlations found in the training data. However, the models might not provide insights into why a specific

prediction was made or what factors influenced it. This lack of interpretability can lead to distrust and skepticism about the decision-making process3.

3 Importance of context: AI systems can struggle to grasp context, including the broader social, cultural, and ethical aspects that might be relevant to a decision-making process. This limitation makes it challenging for AI systems to provide explanations that align with human values and reasoning4.

The lack of transparency and explainability in AI decision-making is a significant concern, particularly in critical domains like healthcare, criminal justice, and finance. Ensuring transparency and explainability is essential to understand and address potential biases, ethical considerations, or errors in AI systems. Researchers and practitioners are actively working to develop techniques and approaches to enhance the transparency and interpretability of AI systems.

AI Limitation: AI systems often struggle with understanding context, meaning, and nuance due to their limited ability to comprehend the complexities of human language, cultural nuances, and situational context. Here are a few key reasons behind this limitation:

1 **Ambiguity in language:** Natural language is intricate, and meanings can vary depending on context, cultural background, idiomatic expressions, and linguistic subtleties. AI systems primarily rely on statistical patterns and predefined rules, which can lead to misinterpretations and misunderstanding of linguistic nuances1.

2 **Lack of common sense reasoning:** AI systems typically lack the comprehensive common-sense knowledge that humans possess. Humans can make inferences, understand implied meanings, and draw upon their general knowledge and experiences to interpret information accurately. AI systems, on the other hand, often struggle with such reasoning, thus impacting their ability to understand context and meaning2.

3 **Cultural and social context:** Understanding context and nuance often requires cultural and social awareness. Many AI systems are trained on specific datasets that may not capture the full spectrum of cultural diversity and context. As a result, the AI may misinterpret

language, cultural references, or social norms, leading to potential errors or misunderstandings3.

4 **Lack of real-world experience:** AI systems are trained on historical data, which may not fully represent the complexities and dynamics of real-world scenarios. They may have limited exposure to the practical aspects of various situations, making it challenging for them to grasp the subtle details and nuances that humans typically understand based on their experiences4.

Addressing these limitations requires advances in natural language processing, machine learning techniques, and incorporating more diverse and inclusive training data. Researchers and practitioners are actively exploring methods to improve AI systems' contextual understanding and language comprehension, aiming to minimize errors and enhance their ability to capture meaning and nuance accurately.

AI Limitation: AI systems have limitations in creativity and imagination compared to humans. While AI has made significant advancements in various domains, it still struggles to match the level of creative output and imagination displayed by humans. Here are some key reasons behind this limitation:

Lack of intuition and emotional understanding: Human creativity often stems from intuition, emotions, and personal experiences. AI systems, on the other hand, rely on algorithms and data patterns, lacking the ability to understand emotions or possess personal experiences12. This limits their ability to generate truly original and emotionally expressive creative outputs.

Inability to think outside predefined algorithms: AI systems operate based on predefined algorithms and patterns learned from training data. They excel at identifying patterns and generating outputs based on those patterns. However, they struggle when asked to produce something truly novel or out of the predefined patterns3. Human creativity, on the other hand, is not limited by such constraints and can generate unique and unexpected ideas.

Contextual and cultural understanding: Creativity often involves understanding rich contextual and cultural cues. Human creators draw upon their knowledge of art, literature, history, and cultural influences to produce creative works. AI systems, although capable of processing large amounts of data, often lack the deep understanding and nuanced interpretation of cultural and contextual elements required for creative output.

Originality and subjective evaluation: Creativity is subjective and involves the evaluation of originality, novelty, and aesthetic value. While AI systems can produce outputs that mimic creative works, they can struggle to replicate the complex and subjective aspects of creativity. Human creators possess the ability to assess and appreciate the subtle nuances that make creative works unique and valuable.

It is important to note that AI systems can still assist and augment human creativity, providing new tools and inspiration for artists and creators. However, they are not yet capable of fully replacing human creativity and imagination.

AI Limitation: dependence on data, which can be limited or biased.

AI systems heavily rely on data for training and making predictions or decisions. This reliance on data can lead to two important limitations: limited data and biased data.

1 **Limited data:** AI systems require large amounts of data to learn patterns and make accurate predictions. However, in some domains, collecting sufficient high-quality data can be challenging. For example, in rare events or niche areas, there may be limited data available to train AI systems, leading to suboptimal performance and limited generalization1.

2 **Biased data:** Another concern is that AI systems may be trained on biased or unrepresentative data, which can lead to biased outcomes. If the training data is biased, the AI system can inadvertently perpetuate and amplify those biases in its predictions or decisions2. Biases can arise from various sources, including historical societal biases, biased data collection processes, or unintentional biases introduced by human labelers3.

Addressing these issues requires careful consideration and proactive steps to mitigate limitations related to limited and biased data:

1 **Data augmentation:** In cases where data is limited, techniques like data augmentation can be employed to augment the existing

dataset, generating additional samples that expand the diversity and variety of training data4.

2 **Unbiased and diverse datasets:** Efforts should be made to create diverse and representative datasets that include examples from various demographic groups, cultural backgrounds, and perspectives. This can help reduce biases and improve the fairness and inclusivity of AI systems5.

3 **Bias detection and mitigation:** It is important to develop robust techniques to detect and mitigate biases in AI systems. This includes regular audits and evaluations of AI models to identify and address potential biases. Additionally, involving diverse teams in AI development can help to minimize unconscious biases in the design and implementation6.

4 **Transparency and accountability:** Organizations and developers should strive for transparency and accountability by documenting the sources and potential limitations of the training data. Openly communicating the strengths and weaknesses of AI systems can foster responsible use and management of these technologies7.

While it is challenging to completely eliminate limitations related to data in AI systems, these strategies can help minimize their impact and ensure more reliable and unbiased AI outcomes

AI Limitation: AI vulnerability to manipulation or hacking can occur due to various factors. One common vulnerability is the reliance on large datasets for training AI models, which can be manipulated or poisoned with false information to influence the AI's decision-making process. Additionally, AI systems can be vulnerable to adversarial attacks, where malicious actors input specially crafted data to deceive the AI into making incorrect predictions or decisions. Furthermore, AI algorithms may have inherent biases based on the data they were trained on, leading to discriminatory outcomes. To mitigate these vulnerabilities, it is essential to implement robust security measures, regularly update AI systems with the latest security patches, and conduct thorough testing to identify and address potential weaknesses.

AI Limitation: Inability to handle unexpected or novel situations.

AI's inability to handle unexpected or novel situations stems from its reliance on pre-existing data and patterns. AI systems are typically trained on historical data and patterns, which means they may struggle to adapt to new or unforeseen circumstances that deviate from the training data. When faced with situations outside their training data, AI systems may provide inaccurate or unreliable results.

To address this limitation, researchers are exploring ways to improve AI's ability to handle unexpected situations. One approach is to incorporate techniques like reinforcement learning, which allows AI systems to learn and adapt in real-time based on feedback from their environment. Additionally, researchers are working on developing AI models that can generalize better across different scenarios and learn from fewer examples.

While AI's inability to handle unexpected situations is a current challenge, ongoing research and advancements in the field aim to enhance AI systems' adaptability and robustness in handling novel scenarios.

AI Limitation: Ethical considerations around the use of AI are crucial due to the potential impact AI systems can have on individuals, society, and the environment. Some key ethical considerations include:

1. **Bias and fairness:** AI systems can inherit biases present in the data used to train them, leading to discriminatory outcomes. It is essential to address bias in AI algorithms to ensure fair and equitable decision-making.

2. **Privacy and data protection:** AI systems often rely on vast amounts of data, raising concerns about privacy and data protection. Safeguarding personal information and ensuring data security are essential considerations when deploying AI technologies.

3. **Accountability and transparency:** AI systems can sometimes operate as "black boxes," making it challenging to understand how they arrive at their decisions. Ensuring transparency in AI algorithms and holding developers accountable for the outcomes of AI systems is crucial.

4. **Safety and reliability:** AI systems must be designed to operate safely and reliably to prevent harm to individuals or society. Robust testing, validation, and monitoring processes are essential to ensure the safety and reliability of AI technologies.

5. **Social impact:** The deployment of AI can have wide-ranging social implications, affecting employment, education, healthcare, and other sectors. It is essential to consider the broader societal impact of AI technologies and work towards maximizing their benefits while minimizing potential harms.

Addressing these ethical considerations requires collaboration among policymakers, industry stakeholders, researchers, and the public to develop guidelines, regulations, and best practices for the responsible use of AI. By prioritizing ethical considerations in the development and deployment of AI technologies, we can harness the potential of AI while upholding ethical standards and values.

AI Limitation: The cost and complexity of developing and implementing AI systems can vary depending on various factors such as the complexity of the AI model, the size and quality of the training data, the expertise of the development team, and the infrastructure required for deployment. Developing AI systems often involves significant costs associated with data collection, preprocessing, model training, and testing. Additionally, hiring skilled AI professionals and acquiring the necessary computing resources can add to the overall development expenses.

The complexity of AI development and implementation also poses challenges, as it requires expertise in machine learning, data science, and software engineering. Building robust AI systems that deliver accurate and reliable results demands careful planning, testing, and optimization throughout the development process. Furthermore, integrating AI systems into existing workflows and systems can introduce additional complexities, such as ensuring compatibility, scalability, and security.

To address the cost and complexity of AI development and implementation, organizations can consider leveraging pre-trained AI models, cloud-based AI services, and automation tools to streamline the development process. Collaborating with experienced AI vendors or consulting firms can also help navigate the complexities of AI implementation and optimize the cost-effectiveness of AI projects.

AI Limitation: AI presents potential risks to job security and the economy as a whole due to automation of tasks traditionally performed by humans. As AI technology advances, there is a concern that certain jobs may become obsolete or significantly reduced in demand, leading to displacement of workers and potential unemployment in those sectors. This can have a ripple effect on the economy, impacting industries, businesses, and individuals.

Furthermore, the rapid adoption of AI could widen the income inequality gap, as those with the skills to work alongside AI technology may benefit while others face job displacement. There is also a risk of job polarization, where high-skilled jobs that require creativity and problem-solving are in demand, while low-skilled jobs are automated, leading to a shrinking middle-class workforce.

To mitigate these risks, policymakers, businesses, and individuals need to proactively address the impact of AI on job security and the economy. This may involve upskilling and reskilling workers to adapt to the changing job market, implementing policies to support displaced workers, and fostering a collaborative approach between humans and AI to create new job opportunities and economic growth.

CHAPTER 23
DISADVANTAGES OF AI

AI requires large amounts of money to set up, maintain, and operate.

Imagine making the first robot that can move packages from one end of the hospital up three flights of stairs and deliver them to a precise location in a hospital. This would require advanced engineering, custom designing and building, and endless testing to determine if it would work properly. This is a significant investment. A company would have to put a lot of money upfront before cashing in on a substantial return. What if it doesn't work or it breaks? You may call a handyperson from Craigslist. If you are lucky, the people who made it would fix it, that is, if they did a good job and had integrity.

Does AI require large amounts of money to set up, maintain, and operate? Chat:

While AI technology can be costly to develop and implement initially, the costs associated with maintaining and operating an AI system can vary depending on the complexity and scale of the project. Data storage, computing power, software updates, and ongoing training may contribute to the overall expenses. However, advancements in AI technology have made it more accessible and cost-effective for businesses and individuals to leverage AI solutions for various tasks. Considering the budget and resources required before investing in an AI system is essential to ensure its successful implementation and operation.

Where's the electricity? Global warming knocked out the power for two months.

What if the power goes out? You may have a solar or wind power system or get your power from the grid. Something happens, and it breaks down. You can fix it, or the company can fix it, but it will get fixed, and we will be right where we were before. Up and going, doing the fast lane, marching full steam ahead using the significant advantages of AI, but is all that for real? It is for a few moments for those in position, the position at the top of the chain of machines that make AI work. Humans made an industrial revolution, and computers and AI can learn and produce better than humans, so what's the problem?

Will AI fail if global warming knocked out the power for two months? Chat:

In a prolonged power outage due to global warming or any other catastrophic event, AI systems that rely on electricity may be affected. Without access to power, AI servers and devices would be unable to operate, disrupting their functionality. However, it is worth noting that some AI systems may have backup power sources or contingency plans to mitigate the impact of power outages. Additionally, advancements in renewable energy sources and energy storage technologies may help ensure the continued operation of AI systems even in challenging circumstances. Organizations must consider resilience and contingency planning when implementing AI solutions to address potential disruptions caused by external factors such as power outages.

Robots do single or a few functions.

After many years of trial and error, humans can do many functions. Once they learn something, it's easy to pick up and do the work you did before. Robots don't work that way. Many are built for a single function. Adding functions increases the time, energy, and money needed to make it work.

Robots, drones, and computer-controlled machines require advanced software and programming.

How do you get a robot to pick up a tomato? After all the mechanical work is done, the computer must tell the robot what to do. The robot must see the tomato, direct an arm to pick it up, and figure out where to put it. Humans may think it's easy, but it requires advanced software to coordinate the sensors, the arms, and the whole machine to work as one unified operation.

Can I live without AI?

A few people will use AI to their advantage but may need to be made aware of the price they pay for it. Many won't have access, and many more won't know what its doing to them.

AI requires a lot of time, energy, and money to set up and maintain. This concentrates power, which is against lower incomes. AI is high on the technology chain; everything below must work. Global warming could wipe out many of the critical components. For computers, robots, and drones, missing one component cam disable the whole machine.

Those who don't use AI, will likely lose advantages. They will lose their jobs, and won't be able to compete with people who use AI. They will be left far behind.

AI will produce extreme inequality unless governments, businesses and society tightly controls it. Some say the solution is to provide a basic income for anyone who wants it. This could work for some people, but may produce similar problems to welfare. People may not be motivated to engage in something beyond the next pay check.

50% of the jobs may be automated

The crushing blow of a lost job takes a considerable toll on people. Computers and robots will replace truck drivers, clerical workers, and retail workers. Even lawyers can't escape the mighty blow. Many boring, repetitive jobs will be done with higher quality and more efficiency and productivity. Unemployment, homelessness, crime, and poverty may increase rapidly. The disruption of American life may end life as we know it for many people. Cities dependent on factory work may deteriorate as incomes vanish, and people can't pay for rent, a vehicle, health insurance, and food.

The long chain of machines

Many machines must work to make a robot, drone, or computer from start to finish. Many operations must work, from mining and processing the metal to designing and building the machine. People get sick, make mistakes, machines break down, and storms hit and knock out production. Robots, drones, and computers may require precious metals, which may be expensive and in short supply. A lot must happen between zero and the final one that links from start to finish. There may be shortcuts at some point, but for now, everything in the long chain must work to get the final result, which may be information. What is the long chain of machines? It's everything between the zero, which means nothing, and the final result you get after you click.

AI can have bugs in it.

Computers, robots, and drones contain software that programs how it functions. These programs can have bugs. The more programs and functions, the more bugs it may contain. For the user, figuring out how to get through the bugs may be an ordeal. Most reputable companies send software patches to fix the problem. These are higher-level machines that do and generate many things.

AI can be idiosyncratic.

People make technology, and when people make something, what they make is a reflection of their ideas, values, and influences. Sometimes, people make mistakes or emphasize one thing more than the other. Choices have to be made. Limitations must be imposed. Something can be done, but only some things. The AI may generate significant errors or malfunction if it isn't rigorously tested.

AI is made for a specific group of people.

Robots are made for people with money to buy them. If you don't have the money, you could use one from someone who has the money. You could also be a part of a collective, which would buy the robot for everyone to use. A local government could buy robots for people in the community, but that's unlikely; mostly, robots are for people with money beyond what is required to meet needs. The number of people who can afford a robot is below 50%. That's a guess. Americans typically have the most disposable income. They can afford clothes washers, dishwashers, and robots that clean floors. I wonder if most people in low-income countries can afford a robot. Ideally, they would not want one.

Robots benefit capitalism more than people.

It's important to remember that people have been living well without robots, computers, and drones for a long time. They learned to acquire the skills needed to advance their quality of life. Without capitalist brainwashing with phones, computers, and television sets, they were able to define what matters and live accordingly. Robots are used to advance capitalism. Yes, consumers benefit, but robots tend to do the work that humans used to do. As the robots take over, many workers lose their jobs, or their wages are cut, and gradually, it becomes apparent that they won't have a job, and their standard of living has been falling. Families fall apart, people get depressed, illnesses increase, they get addicted to painkillers and alcohol, and in a flash, they are gone. For many, a job gives them a sense of self-worth and dignity. They pride themselves on being the person who makes the cash to keep the family going. As workers age, they may lose motivation to educate, train, and develop other skills. Also, older workers have less value in the dominant economy. For those who compete with younger workers, what is going on becomes quite apparent, and that's a big blow to one's self-esteem. If you have to retrain and compete with people with a lot of experience, you must be very clever and outstanding to get a job. Overall, the people who gain the most from robots are at the top. The ones who own the company that owns the robots are the ones who earn the most.

As people depend more on machines, they may reduce their skills

When take-out food becomes familiar, and fewer people learn how to buy and prepare food and clean up, they will have less work, develop fewer skills, and become increasingly dependent on other people and machines. They lose their ability to buy, prepare, and

Robots may do jobs that keep people healthy.

When people work, they get exercise, gain skills, struggle, and suffer. Some people figure out how to do the work faster and more efficiently. All this work can be very satisfying; if you work hard enough, you may sleep well. Many manual labor jobs in construction, manufacturing, and delivery provide exercise at work. What work will people do as AI, robots, and drones do these jobs? Designing, building, and operating AI, robots, and drones will generate many jobs. Some of these may provide exercise, or forward-leaning companies will offer ways to get exercise through work. Many people do whatever they want without required exercise at work, often sitting on the couch way too long.

Can people adapt to AI?

Machines require effort to learn to use them. Some are easier than others. When the technology is new, there are usually more problems, but the fundamental problems will stay. Cars are an example. They started heavy and straightforward and were more accessible to fix. As time passed, they were controlled by a computer, like a CNC machine complex, as companies competed to sell more and more vehicles.

Robots break down

Like cars, trucks, and equipment, robots and technology will break down. Who will fix it? Some companies make things to be repairable, but that's rare. The people who made it could fix it well. They could make robots that repair themselves quickly, which takes another step into the end of humans.

Robots make people more dependent on technology.

As robots, machines, and drones do the work humans do, people may lose their skills to make food, build houses, and care for each other. Do I need to know how to wash my clothes? No, because I'm a privileged person in the US. Is it essential for me to know this? Now, it isn't, but things can change quickly, and I have many other skills, so I could soon learn how to wash my clothes by hand if needed. Most people don't have these advantages, and they may

The infrastructure to build, maintain, and fix robots can't withstand global warming.

Global warming is accelerating as societies head faster and faster into AI and increase production and consumption to feed the capitalist machine. The roads, bridges, and buildings will deteriorate and break faster. People will have to work harder to maintain existing conditions. Large storms will destroy the coastline infrastructure governments spent so much money constructing. To achieve the same standard of living will require more work. So, is it AI to the rescue? Only the most advanced countries will have access to wealthier people. The poor and homeless may be left in the cold. Many in the developing world will be caught between living with the least advantages and the opportunities. To get to these advantages, they have to risk their lives. The road to extinction is getting more significant as the push to consume expands.

The education system and media still need to educate people about the dangers of AI.

You must make an extraordinary effort to realize the impact of AI. Much of the media is focused on selling information to get the attention that sells advertising. Telling people to stop watching TV and start studying, training, and learning other skills to get a job in another field will probably sell poorly, and people who are losing their jobs don't have much money to spend. Advertisers don't want to focus their dollars on people without money. Many people grew up when the assumption was that one person could get a job, have that job for life, and raise a family on that income. Those days are long gone. Now, job after job is outsourced, dumbed down, and automated out of existence. Within ten years, that happened to me twice. Many workers will only go from one dead-end job to another with a skillset compatible with the AI economy. If people have access to media and education, which prepares them for the economic changes ahead, they may be able to adapt without too many negative consequences.

Chat: Will humans become obsolete after AI takes over?

The idea of humans becoming obsolete after AI takes over is a complex and debated topic. While AI technology has the potential to automate specific tasks and processes traditionally performed by humans, it is unlikely that humans will become entirely obsolete. Humans possess unique qualities such as creativity, emotional intelligence, critical thinking, and adaptability that are difficult for AI to replicate. Ethical considerations, decision-making, and complex problem-solving often require human intervention and oversight.

Instead of replacing humans, AI is more likely to augment human capabilities, leading to new opportunities and roles in various industries. Collaboration between humans and AI can increase efficiency, productivity, and innovation. It is essential for society to carefully consider the ethical implications and societal impact of AI adoption to ensure that humans and AI can coexist harmoniously and responsibly.

Chat: Can AI become creative and emotionally intelligent with critical thinking and adaptability skills?

While AI technology has made significant advancements in recent years, there are still limitations to its ability to replicate certain human qualities such as creativity, emotional intelligence, critical thinking, and adaptability. AI systems are designed to process and analyze data to make decisions or perform tasks based on predefined algorithms and patterns. While AI can generate creative outputs such as art, music, or writing, it lacks the depth of human creativity that stems from emotions, experiences, and intuition.

Similarly, emotional intelligence, which involves understanding and managing emotions, empathy, and social skills, is challenging for AI to fully grasp due to the complexity of human emotions and social interactions. Critical thinking and adaptability skills require a deep understanding of context, reasoning, and problem-solving, which AI may only replicate with human intervention.

While AI can assist in certain aspects of creativity, emotional intelligence, critical thinking, and adaptability, it is unlikely to fully replace humans' nuanced and complex capabilities in these areas. Collaborative efforts between humans and AI can leverage the strengths of both to achieve more innovative and effective outcomes.

CHAPTER 24
AI AND GLOBAL WARMING: IS AI SUSTAINABLE?

Global warming increases the frequency and intensity of storms. As weather patterns become less consistent, production becomes less stable. It becomes more challenging to mine and process materials, refine and transport, and machine metals. Hurricanes, tornadoes, fires, floods, and droughts may destroy factories. The conditions for life will deteriorate. When this happens, people can't get to work. These delays drive up the cost of production. Since AI is at the top of the technology chain, it's more vulnerable to global warming than many things; after all, how will anyone make a microchip in the work lab in the basement?

Limiting production to sustainable transport and development has many advantages.

In a way, AI is neutral. AI could be used to advance sustainable transport and development. These limitations accelerate the appropriate use of this technology while avoiding unnecessary development. While this may not be popular, limiting the development and use of AI could bring advantages. With less competition for materials, the cost of production may be similar. The supply of appropriate technology may be more consistent. Also, conflicts may be limited, with fewer people grabbing resources and technology. Less production means less damage to the air, land, water, plants, animals, and people. Because the planet is on the brink of extinction, technology can be limited to what is necessary.

Consumption without limits is freedom, and that's good for capitalism. If you take that away, you are taking away my freedom.

In the US, a common assumption is that as long as I have the money, I should be able to buy and use as much of whatever I want. The government should not limit consumption in any way. More production is an advantage for capitalism, and that's good for people. The environment is not the problem of capitalism. That's a problem for people who choose to live in the wrong place, or because people buy things they don't need, or it's not a problem. Nature is resilient, and a few robots and drones aren't going to destroy your drinking water and food supply. This is a common assumption. We live in a free country, and I can buy and use as much as I want. This idea will accelerate human extinction.

Without limits, humans will go extinct. It's time to limit production and consumption. Can AI do it?

Scientists and people who pay attention know that humans are rapidly destroying the conditions in which people live. Development spreads without limits, and people consume as much as they want without regard for the environment. As the frequency and intensity of storms increase exponentially, the world's capacity to adapt before widespread die-offs of plants, animals, and people is quickly fading away. Given this context, why not limit production and consumption until the earth is stabilized?

AI requires stability

Many robots are stand-alone. They don't need an internet connection to operate. Most require a fuel source, typically electrical, that can be powered differently. A few people have solar panels that generate power for their batteries. These are off the grid and would allow the most independence from the grid. These systems will still have to be maintained. If one or two components are not available, the robot may not function. The robot must be built to be maintained to last. Some are made to break and throw away. It may be possible to create something that can be fixed. Often, this technology changes quickly, so a whole class of technology lasts 5-20 years and is quickly obsolete. Computers are like that. Once a computer is 10-20 years old, there's little, if any, software developed to make it function. Yes, an old skill saw will still function. An old computer system will still function. A lot depends on what level of function is necessary. Keeping up with the latest fashion requires constant updates and changes. It's wasteful and reduces the ecological value of the item. When people use things and throw them out before they are entirely used up, they take a huge environmental toll. The stability of AI depends on large economic systems that continue to produce the components and software for AI to function. When the system is not stable, AI quickly fails. One critical component can make it useless.

Half-broken machines can still function.

Some people make things last. They take good care of the items they use. They don't drop it, don't spill water on it, or hit it with something. They determine when it needs to be fixed and get it serviced before it has no value. Some people, especially in the countryside, are like this. They have machines that are 40 or 50 years old and still working fine. They have gotten their money out of it many times. An article can lose value quickly in the city, where many things are stolen. Something stolen must be sold promptly or stashed so that someone else doesn't steal it, or it's sold in a market. Here, the item has less value than in a store. No one knows the original owner. There's no owner's manual, and some parts may need to be included. After something is stolen, a thief must turn it into money fast. They must get rid of it or risk getting caught with something stolen. In the haste, they have little concern for maintaining its value. People in the market may find it and make it last for a long time. They get it at a discount and want to make it keep going. When global warming wipes out half the village, knowing how to make what you need to live will be helpful.

AI is at the top of the chain of production.

How many tools does it take to grow an onion? It depends on how many onions you want to grow and for whom. Growing an onion with a few tools around the house is possible. Dig some dirt with a shovel, put the seed or onion bulb in the ground, and watch it take off. Or it won't take off; either way, it doesn't take a lot of technology to grow an onion. Of course, if you want to grow a ton of onions, that's another story. Growing onions for yourself or as a hobby doesn't require a lot of complicated machines. Making a robot, a computer, or a drone is another story. AI requires large amounts of materials, skills, and expertise at the highest levels of technology. In 50-100 years, if people are still living on the planet, making robots will seem normal or like what it's like to mass produce a vehicle. Now, those vehicles are made by robots. No one thinks much about it. Similar to a computer or a robot. You don't frequently hear about how complicated it is to achieve this production level.

Who is going to fix my robot when it breaks?

For those with money, the answer is simple: pay someone to do it. Yes, people and probably robots will fix computers and robots. They may be built to be fixed. Some parts of a computer are relatively straightforward. A robot may be more challenging than a computer unless designed to be easily maintained. It may not be such a big issue. Many companies are doing it now. Fixing robots could be a growth industry if robots can't do it. If it breaks, you ship it back to the manufacturer. Most of us are used to products breaking down. Sometimes, it's hard not to wonder if products are made to break down. American automobiles seem that way. Some robots fix themselves.

Where are the parts to fix my robot going to come from?

Most of the components come from manufacturers interested in making it easy to fix, easy to replace, or easy to make it impossible to fix. AI would be made to last more than made to fix and more than made to throw away. Even the best AI electronics, robots, and drones will break. Humans will make sure it will happen. So, where do these components come from? Like computers, the manufacturer will make them the same way the computer is made. Will they be available for someone who wants to fix their AI equipment? Maybe, maybe not. They could be back-ordered. They may not be made anymore, mainly if a small number have been produced.

What will they do with mountains of e-waste containing toxic chemicals?

Imagine how much waste would be generated when 1000 people throw out their robots, drones, or computers. E-waste contains heavy metals, including cadmium, lead, mercury, and arsenic. In a landfill, these chemicals leach into the soil, water, and environment. Some may end up in animals and people. Some toxins enter the food supply, causing and eventually leading to congenital disabilities, cancer, and other health issues. Could robots be built to minimize waste? When the robot is broken, or you don't want it anymore, why not send it back to the manufacturer? They could use what is good, fix what needs to be fixed, and sell it again. If it can't be fixed, it could be melted down, and all the elements could be used again.

Infrastructure can't withstand global warming.

Global warming is accelerating as societies head faster and faster into AI and increase production and consumption to feed the capitalist machine. The roads, bridges, and buildings will deteriorate and break faster. People will have to work harder to maintain existing conditions. Large storms will destroy the coastline infrastructure governments spent so much money constructing. To achieve the same standard of living will require more work. So, is it AI to the rescue? Perhaps, but only the most advanced countries will have full access. Many in the developing world will be caught between living with the most minor advantages and the opportunities in the developed world. To get to these advantages, they have to risk their lives. The road to extinction is getting more prominent as the push to consume expands.

Can AI be used to make the world sustainable? Here's what chatgpt says:

AI can play a significant role in making the world more sustainable. By leveraging AI technologies, we can optimize resource management, reduce waste, and make more informed decisions to address environmental challenges. AI can help in various ways, such as:

1. Energy efficiency: AI algorithms can analyze energy consumption patterns and optimize energy usage in buildings, transportation systems, and industrial processes, leading to reduced energy waste and lower carbon emissions.

2. Environmental monitoring: AI-powered sensors and drones can collect and analyze data on air and water quality, deforestation, wildlife conservation, and climate change, enabling better environmental monitoring and protection.

3. Sustainable agriculture: AI can help farmers optimize crop yields, reduce water usage, and minimize the use of pesticides through precision agriculture techniques, contributing to more sustainable farming practices.

4. Waste management: AI can improve waste sorting and recycling processes, identify opportunities for waste reduction, and optimize waste collection routes, leading to more efficient and sustainable waste management practices.

Overall, AI has the potential to drive innovation and efficiency across various sectors, contributing to a more sustainable world by addressing environmental challenges and promoting responsible resource management.

How can people without money use AI to their advantage?

People without financial resources can still leverage artificial intelligence to their advantage through various means:

1. Open-source AI tools: There are many open-source AI tools and libraries available for free that individuals can use to develop AI applications without incurring costs. Platforms like TensorFlow, PyTorch, and scikit-learn offer many resources for building AI models.

2. Online courses and tutorials: Many online platforms offer free or low-cost classes on AI and machine learning. These resources allow individuals to learn the fundamentals of AI and develop their skills without expensive training programs.

3. Collaborate with others: Joining AI communities and forums can provide opportunities to collaborate with like-minded individuals and share knowledge and resources. People without financial resources can pool their expertise and create impactful AI projects by working together.

4. Utilize cloud services: Some providers offer free tiers or credits for using their AI services, allowing individuals to experiment with AI applications without incurring significant costs. Platforms like Google Cloud, Amazon Web Services, and Microsoft Azure provide access to AI tools and resources for free or at a reduced price.

By leveraging these resources and opportunities, people without financial resources can still benefit from the power of artificial intelligence and enhance their skills in this rapidly evolving field.

How can we use AI to run our buildings without fossil fuels?

To use AI to run buildings without fossil fuels, several strategies can be implemented:

1. Energy Management: AI can optimize energy usage in buildings by analyzing data from sensors, meters, and other sources to identify patterns and trends. AI algorithms can then adjust heating, cooling, lighting, and other systems to minimize energy consumption while maintaining comfort levels.

2. Predictive Maintenance: AI can help predict equipment failures and maintenance needs in buildings, allowing for proactive maintenance to prevent breakdowns and optimize energy efficiency.

3. Renewable Energy Integration: AI can optimize the integration of renewable energy sources like solar panels or wind turbines into building systems. By forecasting energy production and consumption patterns, AI can ensure efficient utilization of renewable energy resources.

4. Demand Response: AI can enable buildings to participate in demand response programs, where energy consumption is adjusted based on grid conditions and pricing. This helps reduce reliance on fossil fuels during peak demand periods.

5. Smart Grid Integration: AI can facilitate communication between buildings and the grid, enabling dynamic energy management and load balancing. This integration allows buildings to adjust energy consumption based on grid conditions and renewable energy availability.

By leveraging AI technologies in these ways, buildings can significantly reduce their reliance on fossil fuels and move towards more sustainable and energy-efficient operations.

How can we use AI to run our vehicles without fossil fuels?

Using AI to run vehicles without fossil fuels involves implementing AI technologies in various aspects of vehicle operation and management to optimize energy efficiency and reduce reliance on traditional fuel sources. Here are some ways AI can be utilized in this context:

1. **Autonomous Driving**: AI-powered autonomous driving systems can optimize routes, speed, and driving behavior to minimize energy consumption and maximize efficiency. By leveraging real-time

data and sensors, AI can make split-second decisions to improve fuel efficiency and reduce emissions.

2. **Predictive Maintenance**: AI can predict and prevent vehicle breakdowns by analyzing data from sensors and monitoring the vehicle's condition in real time. This proactive approach can help reduce fuel consumption by ensuring that vehicles operate at peak performance.

3. **Energy Management**: AI algorithms can optimize energy usage in electric vehicles by managing battery charging and discharging cycles based on driving patterns, traffic conditions, and energy availability. This can extend the vehicle's range and reduce the need for fossil fuels.

4. **Traffic Optimization**: AI-powered traffic management systems can reduce congestion and idling time, which can lead to significant fuel savings. By optimizing traffic flow and reducing stop-and-go driving, AI can help vehicles operate more efficiently.

5. **Vehicle-to-Grid Integration**: AI can facilitate vehicle-to-grid integration, allowing electric vehicles to store and share energy with the grid. This bi-directional energy flow can help balance energy demand and supply, reduce reliance on fossil fuels, and promote renewable energy integration.

By leveraging AI technologies in these ways, it is possible to run vehicles without fossil fuels more efficiently and sustainably, contributing to a cleaner and greener transportation system.

ABOUT THE AUTHOR

Dan Paul grew up in Lake Wobegon fishing, playing sports and building houses. After attending the University of Wisconsin, Madison and working at the linear fusion reactor lab, he explored alternatives to the dominant codes. While staying in coops, communes, and squats around the country, he learned the fundamentals of setting up and operating progressive, radical and sustainable alternatives to the code of the corporation and nation.

ALSO BY DAN PAUL

Philosophy of Action Design and Multiplicity by Dan Paul

Book 1 Auto FreeDesign

Book 2 Workers Health Handbook

Book 3 Save Your Life Prevent Hospital Use

Book 4 Coop OwnersHandbook

Book 5 Eye on AI Meeting NeedsSustainably

Book 6 Travels on the NomadicTerrain

Book 7 Tales of the UrbanShaman

Book 8 Housing in the Danger Zone

Book 9 CoronaTime

Book 10 Philosophy of Design, Action and Multiplicity

Coming soon

Landlords Against Eviction

Auto Free USA

Websites

https://viaradmedia.org

https://autofreedesign.com

http://workershealthhandbook.com

http://sylphu.com